Fungal Pigments

Special Issue Editors

Laurent Dufossé

Yanis Caro

Mireille Fouillaud

MDPI • Basel • Beijing • Wuhan • Barcelona • Belgrade

MDPI

Special Issue Editors
Laurent Dufossé
University of Reunion Island
France

Yanis Caro
University of Reunion Island
France

Mireille Fouillaud
University of Reunion Island
France

Editorial Office
MDPI AG
St. Alban-Anlage 66
Basel, Switzerland

This edition is a reprint of the Special Issue published online in the open access journal *Journal of Fungi* (ISSN 2309-608X) in 2017 (available at: http://www.mdpi.com/journal/jof/special_issues/fungal_pigments).

For citation purposes, cite each article independently as indicated on the article page online and as indicated below:

Lastname, F.M.; Lastname, F.M. Article title. *Journal Name*. **Year**. Article number, page range.

First Edition 2018

ISBN 978-3-03842-787-2 (Pbk)
ISBN 978-3-03842-788-9 (PDF)

Table of Contents

About the Special Issue Editors

Laurent Dufossé has held the position of Professor of Food Science since 2006, at the Reunion Island University, which is located on a volcanic island in the Indian Ocean, near Madagascar and Mauritius. The island is one of France's overseas territories with almost one million inhabitants and the university has 15,000 students. Previously, Professor Dufossé was a researcher and senior lecturer at the Université de Bretagne Occidentale, Quimper, Brittany, France. He attended the University of Burgundy, Dijon, where he received his PhD in Food Science in 1993 and has been involved in the field of biotechnology of food ingredients for more than 28 years. His main research over the last 20 years has focused on microbial production of pigments and studies are mainly devoted to aryl carotenoids, such as isorenieratene, C50 carotenoids, azaphilones and anthraquinones. This research has applications in food science, notably within the cheese industry, the sea salt industry, etc.

Yanis Caro, Assistant Professor, Reunion Island University. Dr. Yanis Caro has held this position since 2010; it is a French university located in the Indian Ocean. Previously, he was a research assistant at CIRAD in Montpellier, France and in Reunion (CIRAD is a French research centre working with developing countries to tackle international agricultural and development issues). Dr. Caro attended the Institut National Polytechnique, Toulouse, where he received his PhD in Food science in 2001. He has been involved in the field of lipid chemistry and biotechnology of natural colorants for more than 15 years. Before joining the university, he was a graduate engineer and technological development advisor at CRITT in Reunion (CRITT is a technological multiservice support centre dedicated to the development of innovations and to the transfer of technologies) and was in charge of food safety innovation projects for Agri-food private enterprises. His main research has focused on production of fungal pigments, biofuels and on extraction and valorization of high-value lipid bioproducts from tropical plants. Studies were mainly devoted to red polyketides pigments from filamentous fungi.

Mireille Fouillaud, Senior Lecturer, at the University of Reunion and a researcher at the Laboratory of Chemistry of Natural Substances and Food Sciences (LCSNSA). Following completion of a Ph.D. in Cell Biology and Microbiology from the University of Aix-Marseille, in 1994, Dr. Fouillaud was recruited to the Faculty of Sciences and Technology of the University of La Reunion. In 2009, she joined the Ecole Supérieure d'Ingénieurs Réunion Océan Indien (ESIROI), in the food engineering department. Dr. Fouillaud teaches biology and microbiology applied to agribusiness. Her main research fields range from the microbial diversity of ecosystems and organisms to the production of metabolites of interest for industries, through biotechnology. Six years ago, she decided to focus on pigmented metabolites obtained from marine-derived filamentous fungi, with potential applications in foods or dyeing industries.

Journal of
Fungi

MDPI

Editorial

Fungal Pigments: Deep into the Rainbow of Colorful Fungi

Laurent Dufossé [1,2,*], Yanis Caro [1,2,*] and Mireille Fouillaud [1,2,*]

[1] Laboratoire de Chimie des Substances Naturelles et des Sciences des Aliments—LCSNSA EA 2212,
    Université de la Réunion, 15 Avenue René Cassin, CS 92003, F-97744 Saint-Denis CEDEX 9,
    Ile de la Réunion, France
[2] Ecole Supérieure d'Ingénieurs Réunion Océan Indien—ESIROI agroalimentaire, 2 Rue Joseph Wetzell,
    F-97490 Sainte-Clotilde, Ile de la Réunion, France
*   Correspondence: laurent.dufosse@univ-reunion.fr (L.D.); yanis.caro@univ-reunion.fr (Y.C.);
    mireille.fouillaud@univ-reunion.fr (M.F.)

Received: 3 August 2017; Accepted: 3 August 2017; Published: 7 August 2017

With the impact of globalization on research trends, the search for healthier life styles, the increasing public demand for natural, organic, and "clean labelled" products, as well as the growing global market for natural colorants in economically fast-growing countries all over the world, filamentous fungi started to be investigated as readily available sources of chemically diverse pigments and colorants. For all of these reasons, this Special Issue of the *Journal of Fungi* highlights exciting new findings, which may pave the way for alternative and/or additional biotechnological processes for industrial applications of fungal pigments and colorants. Eight research papers and one review constitute the journal's final Special Issue.

Our first target when building this project was to welcome papers on the following topics:

The fungal biodiversity from terrestrial and marine origins, bringing new elements about fungi as potential sources of well-known carotenoid pigments (e.g., β-carotene, lycopene) and other specific pigmented polyketide molecules, such as *Monascus* and *Monascus*-like azaphilones, which are yet not known to be biosynthesized by any other organisms such as higher plants. These polyketide pigments also include promising and unexplored hydroxy-anthraquinoid colorants from Ascomycetous species.

The investigation of biosynthetic pathways of the carotenoids and polyketide-derivative colored molecules (i.e., azaphilones, hydroxyanthraquinones, and naphthoquinones) in pigment-producing fungal species.

The description of alternative greener extraction processes of the fungal colored compounds, along with current industrial applications, description of their limits and further opportunities for the use of fungal pigments in beverage, food, pharmaceutical, cosmetic, textile and painting areas.

All these subjects and more are covered by articles published in this Issue:
http://www.mdpi.com/journal/jof/special_issues/fungal_pigments.

*   *Fungal biodiversity from terrestrial and marine origins:*
Production and New Extraction Method of Polyketide Red Pigments Produced by Ascomycetous Fungi from Terrestrial and Marine Habitats by Lebeau J. et al. doi:10.3390/jof3030034.

Biodiversity of Pigmented Fungi Isolated from Marine Environment in La Réunion Island, Indian Ocean: New Resources for Colored Metabolites by Fouillaud et al. doi:10.3390/jof3030036.

Biosynthesis of fungal pigments and ways to increase the efficacy of biosynthetic routes and/or the diversity of the biosynthesized pigments:

Combinatorial Biosynthesis of Novel Multi-Hydroxy Carotenoids in the Red Yeast *Xanthophyllomyces dendrorhous* by Pollmann et al. doi:10.3390/jof3010009.

Carotenoid Biosynthesis in *Fusarium* by Avalos J. et al. doi:10.3390/jof3030039.

Biosynthesis of Astaxanthin as a Main Carotenoid in the Heterobasidiomycetous Yeast *Xanthophyllomyces dendrorhous* by Barredo J.L. et al. doi:10.3390/jof3030044.

In situ microscopic analysis of fungal pigments applied on surfaces:

Microscopic Analysis of Pigments Extracted from Spalting Fungi by Vega Gutierrez S.M. and Robinson, S.C. doi:10.3390/jof3010015.

New modes of extraction of fungal pigments (perstraction, pressurized liquid extraction technique):

Perstraction of Intracellular Pigments through Submerged Fermentation of *Talaromyces* **spp. in a Surfactant Rich Media: A Novel Approach for Enhanced Pigment Recovery** by Morales-Oyervides L. et al. doi:10.3390/jof3030033.

Part of **Production and New Extraction Method of Polyketide Red Pigments Produced by Ascomycetous Fungi from Terrestrial and Marine Habitats** by Lebeau J. et al., with investigation of a pressurized liquid extraction technique. doi:10.3390/jof3030034.

Fine chemical analysis of extracted fungal pigments:

Utilization of High Performance Liquid Chromatography Coupled to Tandem Mass Spectrometry for Characterization of 8-O-methylbostrycoidin Production by Species of the Fungus *Fusarium* by Busman, M. doi:10.3390/jof3030043.

Application of fungal pigments in the industry:

Assessment of the Dyeing Properties of the Pigments Produced by *Talaromyces* spp. by Morales-Oyervides L. et al. doi:10.3390/jof3030038.

We, as Guest Editors, trust all readers of this Special Issue enjoy the contents and we would like to deeply thank all 34 authors who contributed (sorted by their last name), also Prof. Dr. David S. Perlin, Editor-in-Chief of the *Journal of Fungi*, and the editing team at MDPI:

Avalos, Javier
Barredo, Jose L.
Barreiro, Carlos
Bode, Helge B.
Breitenbach, Jürgen
Busman, Mark
Caro, Yanis
Cuet, Pascale
Dufossé, Laurent
Fouillaud, Mireille
García-Estrada, Carlos
Hornero-Méndez, Dámaso

Kosalkova, Katarina
Lebeau, Juliana
Limón, María Carmen
Llorente, Melissa
Magalon, Helene
Méndez-Zavala, Alejandro
Montañez, Julio Cesar
Morales-Oyervides, Lourdes
Oliveira, Jorge
Pardo-Medina, Javier
Parra-Rivero, Obdulia
Petit, Thomas

Pollmann, Hendrik
Robinson, Sara C.
Rodríguez-Ortiz, Roberto
Ruger-Herreros, Macarena
Sandmann, Gerhard
Sousa-Gallagher, Maria
Vega Gutierrez, Sarath M.
Venkatachalam, Mekala
Vinale, Francesco
Wolff, Hendrik

Author Contributions: Laurent Dufossé, Yanis Caro and Mireille Fouillaud wrote the editorial.

Conflicts of Interest: The authors declare no conflict of interest.

Journal of
Fungi

MDPI

Article

Production and New Extraction Method of Polyketide Red Pigments Produced by Ascomycetous Fungi from Terrestrial and Marine Habitats

Juliana Lebeau [1], Mekala Venkatachalam [1], Mireille Fouillaud [1], Thomas Petit [2], Francesco Vinale [3], Laurent Dufossé [1] and Yanis Caro [1,*]

[1] Laboratoire de Chimie des Substances Naturelles et des Sciences des Aliments (LCSNSA), Université de la Réunion, F-97490 Sainte-Clotilde, Ile de la Réunion, France; juliana.lebeau@gmail.com (J.L.); mekalavenkat@gmail.com (M.V.); mireille.fouillaud@univ-reunion.fr (M.F.); laurent.dufosse@univ-reunion.fr (L.D.)

[2] UMR QualiSud, Université de la Réunion, IUT, F-97410 Saint-Pierre, Ile de la Réunion, France; thomas.petit@univ-reunion.fr

[3] Istituto per la Protezione Sostenibile delle Piante (IPSP-CNR) and Dipartimento di Agraria, Università degli Studi di Napoli Federico II, 80055 Portici, NA, Italy; francesco.vinale@ipsp.cnr.it

* Correspondence: yanis.caro@univ-reunion.fr; Tel.: +262-262-483-361

Received: 30 May 2017; Accepted: 23 June 2017; Published: 28 June 2017

Abstract: The use of ascomycetous fungi as pigment producers opens the way to an alternative to synthetic dyes, especially in the red-dye industries, which have very few natural pigment alternatives. The present paper aimed to bio-prospect and screen out 15 selected ascomycetous fungal strains, originating from terrestrial and marine habitats belonging to seven different genera (*Penicillium, Talaromyces, Fusarium, Aspergillus, Trichoderma, Dreschlera,* and *Paecilomyces*). We identified four strains, *Penicillium purpurogenum rubisclerotium, Fusarium oxysporum,* marine strains identified as *Talaromyces* spp., and *Trichoderma atroviride,* as potential red pigment producers. The extraction of the pigments is a crucial step, whereby the qualitative and quantitative compositions of each fungal extract need to be respected for reliable identification, as well as preserving bioactivity. Furthermore, there is a growing demand for more sustainable and cost-effective extraction methods. Therefore, a pressurized liquid extraction technique was carried out in this study, allowing a greener and faster extraction step of the pigments, while preserving their chemical structures and bioactivities in comparison to conventional extraction processes. The protocol was illustrated with the production of pigment extracts from *P. purpurogenum rubisclerotium* and *Talaromyces* spp. Extracts were analyzed by high-performance liquid-chromatography combined with photodiode array-detection (HPLC-DAD) and high-resolution mass spectrometry (UHPLC-HRMS). The more promising strain was the isolate *Talaromyces* spp. of marine origin. The main polyketide pigment produced by this strain has been characterized as *N*-threoninerubropunctamine, a non-toxic red *Monascus*-like azaphilone pigment.

Keywords: red pigment; fungal pigment; *Talaromyces*; *Penicillium*; marine fungi; ascomycetous; *N*-threoninerubropunctamine

1. Introduction

Natural colorants are widely used in the world in many industries such as food, cosmetics, pharmaceuticals, and textiles. The majority of authorized natural food colorants in the market are of either a plant or vegetable origin, and have numerous drawbacks such as instability against light, heat, or adverse pH, and a low water solubility [1]. The dye industry is currently suffering from the cost increase of feedstock associated with the higher demands of eco-friendly pigments for replacing synthetic dyes (like azo dyes). This is even more the case in the red-dye industries, which have no, or very few, natural red pigment alternatives for food processes. For instance, there is a strong need for red colours other than plant-originated anthocyanins, which cannot be used over the whole pH range. As of now, red pigments used in foods are mainly from insects (carmine). Carmine (or carminic acid, cochineal extract) is produced in Peru, Bolivia, Mexico, Chile, and Spain (Canary islands) from the dried bodies of female cochineal insects (*Dactylopius coccus*), primarily grown on *Opuntia* cacti [2]. Carmine is considered as one of the most stable natural food red colorants in terms of light and heat. Between 2004 and 2009, a 76% increase in new European food product launches listing carmine as an ingredient was observed, an increase also linked to the consequences of the "Southampton six" study, which promoted a warning for child hyperactivity related to the occurrence of six artificial colorants in food, including three sulphonated mono azo red dyes (E122 carmoisine/azorubine, E124 Ponceau 4R, and E129 Allura Red AC). However, carmine holds ethical issues for some social groups, and another drawback of carmine products is that from a stable level of 15 USD per kg, it surged in 2010–2011 up to 120 USD per kg and decreased again to 15 USD per kg. As a conclusion, "Dr Jekyll's" (positive) aspect of carmine is its excellent stability in food formulations, whereas the "Mr Hyde" (negative) ones are: (i) it cannot be used by vegans-vegetarians-kosher-halal, (ii) its price versatility, and (iii) allergenicity in some cases [3]. The world's largest food color company, Chr. Hansen, which sources one third of global carmine production, decided in 2011 to explore whether it would be commercially viable to produce carmine with a controlled fermentation process (proof of concept test).

Thus, there is an increasing interest from the academic world and industrial sector about the readily available natural sources of red pigments. Among non-conventional sources, ascomycetous fungi are known to produce an extraordinary range of red polyketide pigments that are often more stable and soluble than plant pigments [4–6]. So, fungal red polyketides, such as azaphilone, naphtoquinone, and hydroxyanthraquinone red compounds, are most promising in this respect, even if unusual microbial red carotenoids should be investigated. The development of such a fungal-based pigments industry and its sustainability rely on the selection of adequate strains regarding the three following parameters: (i) profitable yields, (ii) pigment purity and stability, (iii) and the total absence of toxic compounds in the fungal pigment extract. Furthermore, fungal pigments are of interest due to the broad spectrum of their biological activities and their potential applications in designing new pharmaceutical products [7].

Nowadays, some fermentative natural colorants from filamentous fungi like *Blakeslea trispora*, *Ashbya gossypii*, *Penicillium oxalicum*, and *Monascus* sp., are available for replacing the yellow, orange, and red synthetic dyes [6–9]. Over the past five years, very few reports have been published on the *Monascus*-like azaphilone red pigments produced by non-mycotoxigenic strains of *Talaromyces* species [6–9]. In the literature, this biosynthetic potential has been linked to species such as *Talaromyces purpurogenus*, *T. albobiverticillius*, *T. marneffei*, and *T. minioluteus*, often known under their previous *Penicillium* names. For example, in 2012, a European patent was granted for a submerged cultivation method for some of the non-mycotoxigenic strains of *Talaromyces* sp., whereby the concentration of pigments was significantly enhanced, with the polyketide azaphilone purple pigment PP-V [(10Z)-12-carboxyl-monascorubramine] being the major compound [10–12]. N-glutarylmonascorubramine and N-glutarylrubropunctamine were the water-soluble *Monascus*-like polyketide azaphilone red pigments discovered in the extracellular pigment extract obtained from the liquid medium of *Penicillium purpurogenum* [13]. Recently, Frisvad et al. [14] concluded that the isolate of *T. atroroseus* sp. nov., which produces *Monascus*-like azaphilone red pigments and mitorubrins, without being accompanied by mycotoxin synthesis, can be used industrially for red pigment production (patent application EP2262862 B1 [12]). However, they indicated that isolates identified as *T. purpurogenus* may not be recommended for the industrial production of red pigments due to their potential coproduction of mycotoxins, such as rubratoxin A and B, and luteoskyrin, in addition to potential toxic extrolites, such as spiculisporic acid and rugulovasine A and B.

Few reports have been published on the following polyketide naphthoquinone red pigments produced by *Fusarium* species: aurofusarin in *Fusarium graminearum* [15] and bikaverin and its minor coproduct nor-bikaverin in *Fusarium fujikuroi* [6,16]. Along similar lines, some species of the genus *Aspergillus* were found to produce known polyketide hydroxyantraquinone red pigments, such as erythroglaucin, catenarin, and rubrocristin [2,6,17]. Some strains of *Trichoderma* such as *T. aureoviride*, *T. harzianum*, and *T. polysporum* are found to produce the hydroxyanthraquinone orange-red pigment chrysophanol [6]. The hydroxyanthraquinone red pigments catenarin and erythroglaucin have also been isolated from cultures of strains among *Drechslera* species and from a culture of *Curvularia lunata* [6].

The present paper aimed to bio-prospect and screen out 15 selected ascomycetous fungal strains, belonging to seven different genera (*Penicillium*, *Talaromyces*, *Fusarium*, *Aspergillus*, *Trichoderma*, *Dreschlera*, and *Paecilomyces*) originating from terrestrial and marine habitats. Recent literature abundantly reports the interest in marine microorganisms with respect to the production of new molecules and, among them, new pigments [7]. The biotechnological properties of the 15 strains for the production of extracellular water-soluble pigments and intracellular polyketide pigments were investigated in submerged cultures.

The choice of extraction protocol is crucial, as the extraction solvents and conditions can drastically influence the final composition, quality, and efficiency of the process. Indeed, extended extraction times, and the exposure to organic solvents and a higher temperature, can result in a tremendous loss of bioactive substances due to hydrolysis, oxygen- and light-oxidation, as well as ionization. In order to preserve as much as possible of the qualitative and quantitative compositions of the pigmented molecules, the use of a recent extraction method, known as Pressurized Liquid Extraction, was investigated, resulting in the development of a six-stage pressurized liquid extraction protocol (PLE) for advanced mycelial pigment extraction. PLE has been mostly applied on environmental samples (recovery) [18], as well as food and biological samples [19], as an analytical method. Only a few applications have been reported on the extraction of bioactive phyto-compounds. Thus, to our knowledge, such methods, which use less and non-toxic solvents, have not been used on fungal matrixes thus far.

2. Materials and Methods

2.1. Fungal Strains

Eleven fungal strains used in this study originating from terrestrial environments were bought from the fungal culture collection of the Museum d'Histoire Naturelle de Paris (Paris, France): *Penicillium purpurogenum* LCP4890, *P. purpurogenum rubisclerotium* LCP4464, *P. erythromellis* LCP3684, *P. oxalicum* LCP4158, *Fusarium oxysporum* LCP531, *Aspergillus repens* LCP5511, *Paecilomyces farinosus* LCP3391, *Trichoderma harzianum* LCP3404, *T. polysporum* LCP3531, and *Dreschlera cynodontis* LCP2226. *Trichoderma harzianum* strain T22 is a commercial biological control strain. The four fungal isolates of marine origin investigated in this study and identified as *Talaromyces* spp. (code: 305_70), *Talaromyces verruculosus* (code: PA9), *Trichoderma atroviride* (code: 305_55), and *Aspergillus sydowii* (code: B34) were isolated by Mireille Fouillaud from samples collected in the back reef-flat and on the external slope of the coral reef on the west coast of La Reunion island. The fungal collection was stored at −80 °C at the LCSNSA laboratory (Reunion island).

2.2. Fermentation and Biomass Production

For inoculum preparation, 0.15 g of conidia and mycelium mixture was sampled from a seven days-old preculture on a potato dextrose agar (PDA) plate, and transferred into a microcentrifuge tube containing 1 mL of nutrient broth supplemented with 0.05 g·L^{-1} of Tween® 80 (Sigma-Aldrich Co, Saint Louis, MO, USA). The mycelium was crushed and the suspension was used to inoculate 250-mL flasks containing 100 mL of liquid media: (i) potato dextrose broth medium (PDB: composed of 4 g·L^{-1} potato infusion solids and 20 g·L^{-1} glucose; Sigma-Aldrich); (ii) defined minimal dextrose broth medium (DMD: composed of 1 g·L^{-1} ammonium sulfate, 30 g·L^{-1} glucose, 0.5 g·L^{-1} MgSO$_4$, 1.4 g·L^{-1}

K_2HPO_4, 0.6 $g \cdot L^{-1}$ KH_2PO_4, 0.8 $mg \cdot L^{-1}$ $ZnSO_4$, 0.8 $mg \cdot L^{-1}$ $FeSO_4$, 0.8 $mg \cdot L^{-1}$ $CuSO_4$, 0.8 $mg \cdot L^{-1}$ NaH_2PO_4 and 0.4 $mg \cdot L^{-1}$ $MnSO_4$; Fisher Scientific UK Limited, Loughborough, Leicestershire LE, UK) based on Velmurugan et al. [20]; and (iii) yeast casamino dextrose broth (YCD: composed of 1 $g \cdot L^{-1}$ yeast extract (Becton, Dickinson and Co., Sparks, MD, USA), 5 $g \cdot L^{-1}$ casamino acids (BD Bacto), 20 $g \cdot L^{-1}$ glucose, 5 $g \cdot L^{-1}$ sodium chloride and 1 $g \cdot L^{-1}$ KH_2PO_4 (Fisher Scientific UK Limited, Loughborough, Leicestershire LE, UK) based on Guyomarc'h et al. [21]. The pH was adjusted to 6.0. Flasks were incubated at 26 °C and agitated at 150 rpm. For cultures performed in total darkness, flasks were wrapped in aluminum foil. After seven days of fermentation, all the contents of each flask were collected and centrifuged at 10,000 rpm for 10 min; the resulting supernatant was filtered through a Whatman filter paper (GF/C) at a reduced pressure using a Büchner funnel to obtain the culture filtrate. The mycelial biomass was washed with deionized water. After freezing at −84 °C in an ultra-low-temperature freezer (Sanyo, Guangzhou, China) for at least 2 h, the samples were quickly transferred to a LABCONCO FreeZone 2.5 lyophilizer (LABCONCO, Kansas City, MO, USA) and lyophilized for 24 h. During freezing, the condenser temperature and vacuum pressure were maintained at −47 °C and 200 mbar, respectively. Then, dried cells were weighed to estimate the biomass. All experiments were conducted in duplicate.

2.3. Quantitative Colorimetric Analysis of Extracellular Extracts

The colorimetric characterization of extracellular extracts was assessed from the pigmented culture filtrate after seven days of cultivation. Measurements were performed in the CIE L*a*b* (L*a*b* colorimetric system of the Commission Internationale de l'Eclairage) using a spectrocolorimeter CM 3500 with the SpectraMagic™ software v1.9 (Konica Minolta, Mahwah, NJ, USA). The so-called CIELab colorimetric system is based on the fact that light reflected from any colored surface can be visually matched by an additive mixture of the three primary colors: red, green, and blue [22,23]. To characterize a color in the CIE L*a*b* color system, three colorimetric coordinates are obtained from the spectrocolorimeter. L* defines the lightness (ranges from 0% to 100%, dark to light), a* value indicates the red/green value (from −60 to +60, green to red), and b* value denotes the blue/yellow value (from −60 to +60, blue to yellow). The attributes of color, C* and h°, describe the chroma (vividness or dullness) and the hue angle (or tone) of the color, respectively. The value of chroma C* is 0 at the center and increases according to the distance from the center. The hue angle h° is defined as starting at the +a* axis and is expressed in degrees: 0° would be +a* (red), 90° would be +b* (yellow), 180° would be −a* (green), and 270° would be −b* (blue). Hue values correspond to the angle of the a*/b* coordinates of the points. The "Y red chroma" value used in this study to link the positive a*-value (red color) and the others coordinates and attributes of the color of the pigment extract (such as b*, C* and h* values) was calculated as follow:

$$\text{"Y red chroma" value} = f(a^* \text{ value}) = b^* \text{ value} \times C^* \text{ chroma value} \times 1/h^* \tag{1}$$

2.4. UV-Visible Spectrophotometry and Extracellular Polyketide Metabolites Quantification

The culture filtrate was diluted in deionized water with a dilution factor (d) ranging from four to 10 with respect to the concentration of extrolites in the filtrate. The solution was used to investigate the sample absorption profile in the wavelength range of 200–800 nm. Then, the concentrations of polyketide metabolites in the culture filtrate were determined through the absorption values read at 276 nm using a UV-visible spectrophotometer (UV-1800, Shimadzu Corporation, Tokyo, Japan). Polyketide red carmine aqueous solutions were used as standards, and their maximum absorption values experimentally measured at 276 nm were taken as the reference for polyketide concentration determination (Figure S1). Carmine (hydrosoluble carmine in powder, kindly provided by Pronex, Lima, Peru) used as the standard was dissolved in pure water. The diluted uninoculated liquid broth (that is PDB, DMD, or YCD broth) was used as a blank before any absorbance measurements were taken. The coefficient of proportionality (ε), which links the absorbance at 276 nm with the extracellular polyketide metabolites concentration, was obtained by linear regression (Figure S1). The concentration of extracellular polyketide metabolites in culture filtrate was expressed as milli-equivalents of polyketide carmine per liter of liquid broth.

2.5. New Extraction Method of Intracellular Polyketide Pigments

The extraction of intracellular polyketide pigments from the mycelial biomass was performed using a new pressurized liquid extraction (PLE) process. The weighed sample (lyophilized biomass) was transferred to a 10-mL stainless steel extraction cell equipped with two cellulose filters on the bottom and containing glass balls (diam. 0.25–0.50 mm). PLE extraction was performed on a Dionex ASE system (ASE^{TM} 350 apparatus, Dionex, Germering, Germany). The ASE conditions were: temperature: 90 °C, pressure: 1500 psi, heating time: 5 min, static time: 18 min, flush: 100%, and purge: 5 min. The lyophilized biomass was subjected to a six-stage liquid solvent extraction under pressure as an attempt to entirely extract the intracellular pigments from the mycelium. The following six-stage PLE sequence was performed: purified water was used as the first extraction solvent, followed by 50% methanol, then 50% ethanol, >99.9% methanol, and MeOH:EtOH (1/1, *v/v*), and then, mycelium was depleted with >99.9% ethanol as the extraction solvent (Figure S2). The sequence of solvents was set to show a decreasing polarity profile. Solvents (methanol and ethanol, 99.9%-HPLC quality) were obtained from Carlo Erba (Val de Reuil, France). Purified water was obtained from a Milli-Q system (EMD Millipore Co., Billerica, MA, USA). A part of each color extract was used for chromatographic analysis and the rest was used for absorbance analysis.

2.6. Absorbance and HPLC-DAD Analyses

Each intracellular extract obtained in a collection bottle at the end of the six-stage pressurized liquid solvent extraction sequence was filtered onto a 0.2-μm poresized hydrophylic Millex-GV membrane (Millipore, Carrigtwohill, Ireland) and stored at −20 °C in amber glass vials (2 mL) with Teflon-lined caps, until further analysis. The total polyketide secondary metabolite content extracted from the mycelial biomass was first analyzed by measuring the absorbance of each extract by spectral analysis at 276 nm. The results obtained are expressed in terms of meqv. of carmine per g dry cell mass, a value proportional to the polyketide metabolite concentration extracted from the mycelia. Then, reverse phase high-resolution liquid-chromatography combined with photodiode array-detection (HPLC-DAD) analysis was performed on each extract (25 μL injection) on a Dionex HPLC-DAD system (Ultimate 3000 apparatus, Dionex, Germering, Germany) and using a Hypersil Gold^{TM} column (150 mm × 2.1 mm i.d., 5 μm; Thermo Scientific Inc., Waltham, MA, USA) maintained at 30 °C. The HPLC-DAD system was operated using a (A) purified water-(B) acetonitrile-(C) aqueous formic acid 1% (*v/v*) solution gradient system starting from a ratio of 45%(A)–45%(B)–10%(C) for 8 min and increasing to 95%(B)–5%(C) in 55 min, at which point it was maintained for 20 min. The flow rate was 0.4 mL·min^{-1}. Monitoring, data recording, and processing were led with the Chromeleon v.6.80 software (Dionex). Solvents (acetonitrile and methanol, 99.9%-HPLC quality) and formic acid (purity 99%) were obtained from Carlo Erba (Val de Reuil, France).

2.7. UHPLC-HRMS Analyses

The intracellular extracts were analyzed by ultra high performance liquid chromatography high-resolution mass spectrometry (UHPLC-HRMS) according to Klitgaard et al. [24]. Liquid chromatography was performed on an Agilent 1290 Infinity LC system with a DAD detector, coupled to an Agilent 6550 iFunnel Q-TOF with an electrospray ionization source (Agilent Technologies, Santa Clara, CA, USA). The separation was performed on a 2.1 mm × 250 mm, 2.7 μm Poroshell 120 Phenyl-Hexyl column (Agilent) at 60 °C with a water-acetonitrile gradient (both buffered with 20 mM formic acid) going from 10 % (*v/v*) to 100 % acetonitrile in 15 min, followed by 3 min with 100 % acetonitrile. The flow rate was kept constant at 0.35 mL/min throughout the run. The injection volume, depending on the sample concentration, typically varied between 0.1 and 1 μL. Mass spectra were recorded as centroid data for m/z 85–1700 in MS mode and m/z 30–1700 in MS/MS mode, with an acquisition rate of 10 spectra/s. The lock mass solution in 95% acetonitrile was infused in the second sprayer using an extra LC pump at a flow of 10–50 μL/min, and the solution contained 1 μM tributyle amine (Sigma-Aldrich), 10 μM Hexakis(2,2,3,3-tetrafluoropropoxy) phosphazene (Apollo Scientific Ltd., Cheshire, UK), and 1 μM trifluoroacetic acid (Sigma-Aldrich) as lock masses.

3. Results

3.1. Biomass and Polyketide Extrolites Production Capacities under Various Growth Conditions

Our results indicated that the 15 fungal strains investigated in the present study could be divided into two main color categories. The reddish polyketide pigment producers include the four following strains, ranked from the darker to the paler red-pigment producers: *P. purpurogenum rubisclerotium*, the local marine isolate identified as *Talaromyces* spp., followed by the strain *F. oxysporum*, and the marine isolate *Trichoderma atroviride*. The yellowish pigment producers encompass the following strains: *P. erythromellis*, *P. oxalicum*, *Talaromyces verruculosus*, *Trichoderma harzianum*, *Paecilomyces farinosus*, *Aspergillus repens*, and *A. sydowii* (Figure S3). The dried biomass concentration produced by the 15 ascomycetous strains is reported in Figure 1. Three different culture broths were investigated: (i) defined minimal dextrose medium (DMD), (ii) potato dextrose broth medium (PDB), and (iii) yeast casamino dextrose broth (YCD).

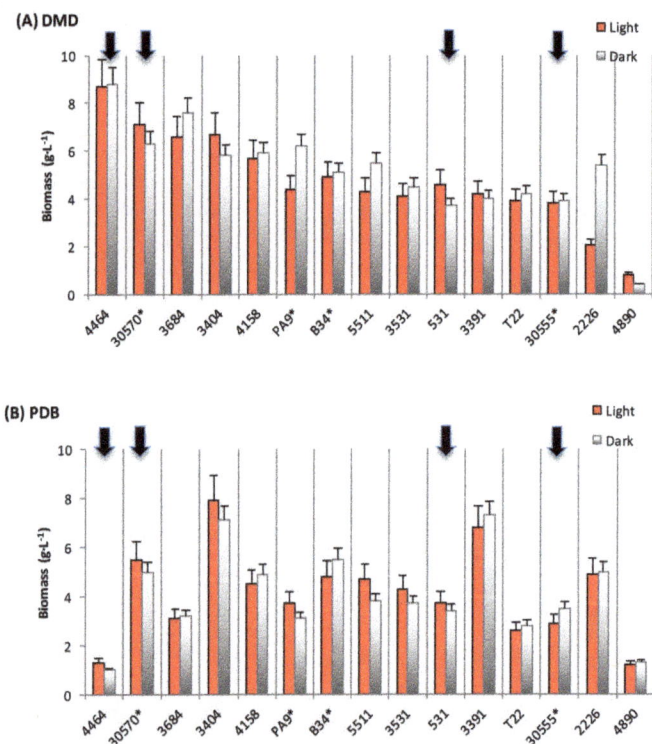

Figure 1. *Cont.*

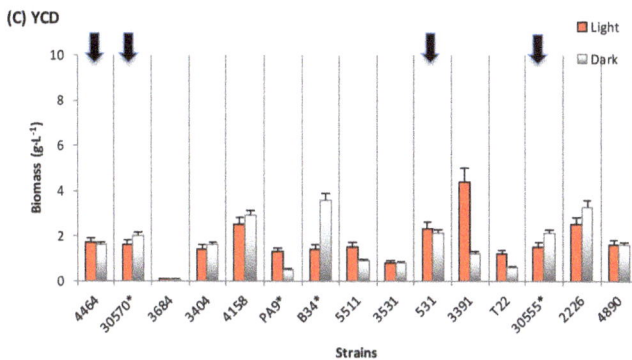

Figure 1. Fungal biomass production (mean in g·L^{-1} ± s.d.) in a submerged culture of the ascomycetous fungi. (**A**) Biomass production obtained in DMD (Defined Minimal Dextrose broth) submerged culture; (**B**) in PDB (Potato Dextrose Broth) submerged culture; (**C**) in YCD (Yeast Casamino Dextrose broth) submerged culture; Culture conditions: under illumination (❙: red) and in the dark (❙: grey); s.d.: standard deviation; Strain identification: 4464 = ➊ *Penicillium purpurogenum rubisclerotium*; 30570 = ➋ *Talaromyces* spp. *(marine isolate)**; 531 = ➌ *Fusarium oxysporum*; 30555 = ➍ *Trichoderma atroviride**; 4890= *Penicillium purpurogenum*; 2226 = *Dreschlera cynodontis*; 3684 = *Penicillium erythromellis*; T22 = *Trichoderma harzianum*; 4158 = *Penicillium oxalicum*; 5511= *Aspergillus repens*; PA9= *Talaromyces verruculosus**; 3404= *Trichoderma harzianum*; 3391= *Paecylomyces farinosus*; B34 = *Aspergillus sydowii**; 3531 = *Trichoderma polysporum*; (* strains collected from marine biotopes of La Reunion island's reef flat).

Our results indicated that the DMD medium (Figure 1A) is the more favorable liquid broth for biomass production. Six fungal strains producing more than 5.5 g·L^{-1} of dry biomass were observed: *P. purpurogenum rubisclerotium*, i.e., the first reddish pigment producer described above, which produced 8.4 and 8.5 g·L^{-1} in the light and in the dark culture, respectively; followed by the marine isolate *Talaromyces* spp. (the second reddish pigments producer: 7.1 and 6.2 g·L^{-1} of dry biomass); and four other strains, i.e., *P. erythromellis* (6.5 and 7.5 g·L^{-1}), *Trichoderma harzianum* (6.7 and 5.8 g·L^{-1}), *P. oxalicum* LCP4158 (5.7 and 5.9 g·L^{-1}), and *T. verruculosus* (4.4 and 6.1 g·L^{-1}). The two other reddish pigment producers, i.e., the strain *F. oxysporum* and the marine isolate *T. atroviride*, produced less than 4.5 g·L^{-1} of biomass in this DMD medium. In PDB medium (Figure 1B), considering the four fungal strains identified above as potential reddish pigment producers, only the marine strain *Talaromyces* spp. produced more than 5.5 g·L^{-1} of dry biomass. Surprisingly, whereas the strain *P. purpurogenum rubisclerotium* showed the highest biomass productions in the DMD medium, it was the less productive strain in PDB medium (1.3 and 1.0 g dry biomass·L^{-1} in the light and in the dark culture, respectively). In YCD medium (Figure 1C), the biomass contents remained relatively lower (<2 g·L^{-1}) for most of the cultured fungal strains.

Figure 2 presents an estimation of the volumetric productions of polyketide extrolites secreted by each strain in the culture filtrate. The concentrations were expressed as milli-equivalents of carmine per liter of liquid broth. The color of the "light" histograms of Figure 2 indicates the shade of the culture filtrate after seven days of fermentation: reddish and yellowish shades were mainly observed, but an orange shade (obtained for *A. repens* cultivated in YCD under light conditions), pink shade (*T. verruculosus* grown in DMD in darkness, or *P. purpurogenum* grown in PDB and YCD), brown shade (*Dreschlera cynodontis* grown in DMD and YCD broths), and green shade (*Trichoderma polysporum* cultivated in PDB and DMD broths) were also noticed (S3). The lowest extracellular productivities (<200 meqv g·L^{-1}) were observed on "minimal" DMD medium for all strains investigated in this study (Figure 2A). Thus, the presence of a simple source of carbon (30 g·L^{-1} of glucose) and simple source of nitrogen (ammonium sulfate) in this DMD medium favored the production of mycelial biomass to the detriment of the production and excretion of extracellular diffusible polyketides. In contrast, the best polyketide extrolite productions were obtained in PDB liquid culture medium, reaching an average extracellular volumetric production ranging from 135 to 1118 meqv·L^{-1} for eight strains out of the 15 (Figure 2B). For example, the highest extrolite production was obtained in culture filtrate of the strain *P. purpurogenum rubisclerotium* grown in PDB liquid broth (1118 and 685 meqv·L^{-1} in the light and dark culture, respectively), while its biomass concentration was very low in this medium (only 1.0–1.3 g·L^{-1} of dry biomass). The marine isolate *Talaromyces* spp. presents a different behavior towards the medium nutrient composition. It simultaneously showed a relatively high mycelial biomass content (above 5.5 g·L^{-1} of dry biomass) in PDB and a strong content in extracellular polyketide-derived metabolites (278 and 266 meqv·L^{-1} in the light and dark culture, respectively). The two other strains belonging to the reddish pigment producers, i.e., *F. oxysporum* and *T. atroviride* (marine isolate), presented extracellular polyketide productivities of 166 and 373 meqv·L^{-1}, respectively, but also low contents in the mycelial biomass in submerged cultures in PDB (< 3.5 g·L^{-1} of dry biomass). The YCD medium that also contained a complex source of nitrogen (f.i. amino acids and proteins from yeast extract) had an intermediate position regarding extracellular polyketide extrolite production, exhibiting very heterogeneous productivities (Figure 2C).

The culture filtrates were then analyzed by quantitative colorimetric analysis in the CIE L*a*b* colorimetric system. It is worth noticing that for colored culture filtrates, the greater the positive red a*-value is, the greater the positive b*-value, the greater the positive C*-chroma, and the smaller the h*-value. Thus, a "Y chroma" value has been calculated in this study to link the positive a*-value (red color) and the other b*, C*, and h* values of the colored culture filtrates. The results are presented in Figure 3. In PDB medium with both complex sources of carbon and nitrogen, the four fungal strains belonging to the reddish pigment producers could produce diffusible polyketide red pigments: the culture filtrates exhibited an intense purple, red, or orange-red color, in the darkness, as well as under light conditions (Figure 3A), which is consistent with visual observations of the culture filtrate and the volumetric production of extracellular pigments noticed in these fungal cultures grown in PDB. The highest positive a*-value and positive "Y chroma" value were obtained for the culture filtrate of the marine isolate *Talaromyces* spp. (purple-red color; a*-value up to +63.8) and the strain *P. purpurogenum rubisclerotium* (purple-red color; a*-value up to +51.5). The two other strains, *F. oxysporum* and *T. atroviride*, exhibited lower positive a*-values: up to +35.2 (red color) and +18.7 (orange-red color), respectively.

Furthermore, it is worth noticing that in the "minimal" DMD liquid broth (with only glucose and ammonium sulfate, combined with some metal ions and inorganic salts), only the marine isolate *Talaromyces* spp. and the strain *F. oxysporum* can produce water-soluble reddish pigments: with a corresponding red a*-value from +10 to +20 (Figure 3B). The strain *P. purpurogenum rubisclerotium* grown in this DMD broth produced pinkish pigments exclusively under light conditions, and a pale yellowish pigmentation was noticed for the marine isolate *T. atroviride* grown in DMD. The results confirmed that the same fungal strain did not show the same metabolism of extracellular polyketide pigments in submerged cultures with respect to the medium composition used. Finally, in YCD medium (with glucose, yeast extract, and casamino acids as the complex nitrogen source), only one strain, that is *P. purpurogenum rubisclerotium*, can produce reddish pigments in liquid broth (red a*-value of +42.9 and +43.5 for light and dark culture, respectively) (Figure 3C). Interestingly, the marine isolate *T. atroviride* cultivated in YCD broth produced an orange-red pigmentation exclusively under light conditions, indicating that the production of orange-red pigments is a light-inducible metabolism.

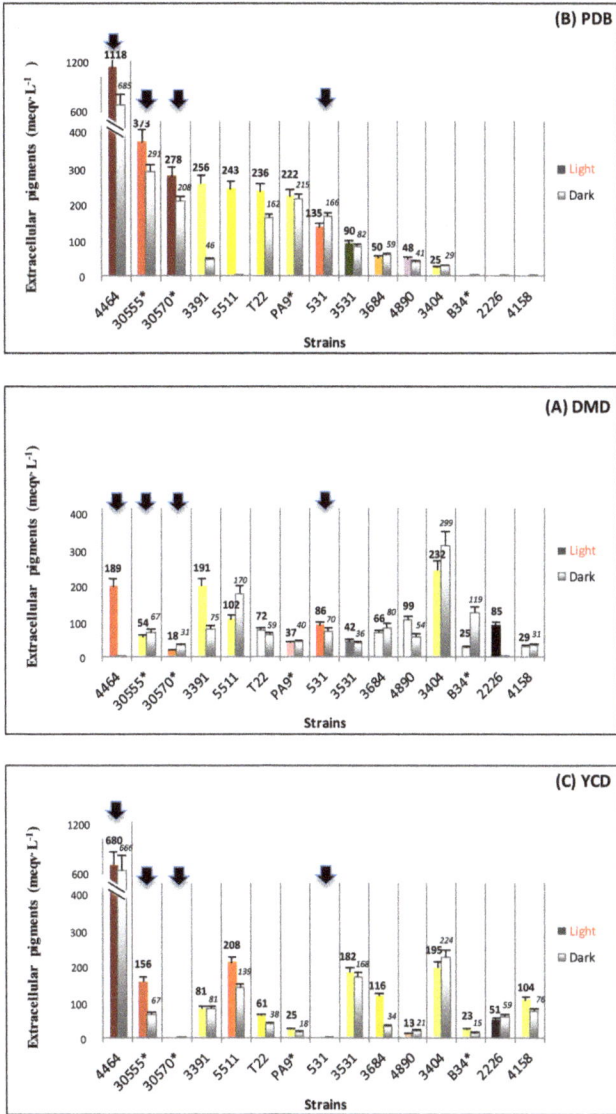

Figure 2. Volumetric production of extracellular pigments (mean in meqv. carmine pigment per liter ± s.d.) by the ascomycetous fungi in submerged culture. (**A**) Volumic production of extracellular pigments obtained in DMD (Defined Minimal Dextrose broth) submerged culture; (**B**) in PDB (Potato Dextrose Broth) submerged culture; (**C**) in YCD (Yeast Casamino Dextrose broth) submerged culture; Culture conditions: in the dark (grey) and under illumination (in color; color in figure for each strain indicates the shade of the fungal culture filtrate after 7 days of fermentation); s.d.: standard deviation; ❶ *Penicillium purpurogenum rubisclerotium* (4464); ❷ *Talaromyces spp.* (marine isolate 30570); ❸ *Fusarium oxysporum* (531); ❹ *Trichoderma atroviride* (marine isolate 30555). For other strain identification (numbers in horizontal axis), see Figure 1. The yield of extracellular pigments was estimated according to a calibration curve of standard carmine solution by measuring the absorbance at 276 nm (λ_{max} of carmine) of the colored culture filtrate (after blank subtraction).

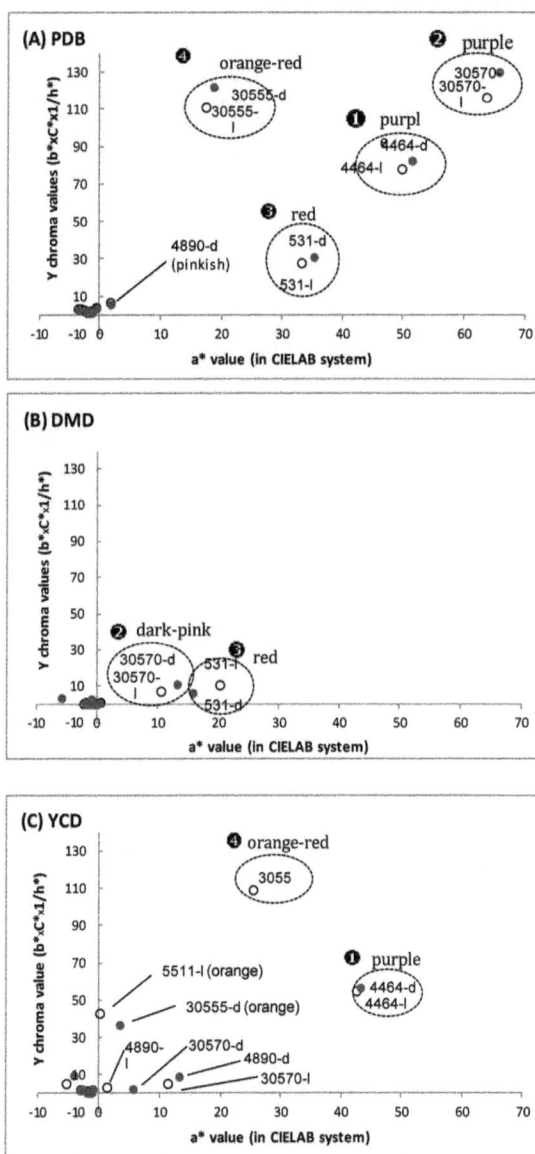

Figure 3. Color coordinates projected in the CIE L*a*b* colorimetric system of the culture filtrates after seven days of cultivation of the ascomycetous fungi. (**A**) Color coordinates of extracellular pigment extracts obtained in PDB (Potato Dextrose Broth) submerged culture; (**B**) in DMD (Defined Minimal Dextrose broth) submerged culture; (**C**) in YCD (Yeast Casamino Dextrose broth) submerged culture; Culture conditions: under illumination (-l), and darkness (-d). Strain identification: ❶ *Penicillium purpurogenum rubisclerotium* (4464); ❷ *Talaromyces spp.* (marine isolate 30570); ❸ *Fusarium oxysporum* (531); ❹ *Trichoderma atroviride* (marine isolate 30555); *Penicillium purpurogenum* (4890).

3.2. New Extraction Procedure and Nature of the Polyketide Red Pigments Produced by Talaromyces spp. Marine Isolate and P. purpurogenum rubisclerotium Terrestrial Isolate

In order to evaluate the intracellular production and the nature of the red pigments produced by the *Talaromyces* spp. marine isolate and *P. purpurogenum rubisclerotium* terrestrial isolate, a series of six sequential pressurized liquid extractions (PLE) were realized. The sequence of solvents was set to show a decreasing polarity profile, therefore allowing a refined isolation of different pigment molecules, depending on their polarity profiles. The total polyketide secondary metabolite contents extracted from the mycelial biomass, analyzed by measuring the absorbance of all the extracts by spectral analysis at 276 nm and expressed in terms of meqv. of polyketide carmine per g dry cell mass, are shown in Table 1. Thus, the results confirmed that the amount of polyketide pigments produced extra- versus intracellularly varies with different cultivation factors and the choice of fungal strain. The DMD medium favors the production of mycelial biomass, whereas the nutrient composition of the PDB medium favors both the intracellular production of polyketide pigments (up to 69.8 and 116.0 meqv·g^{-1} dry biomass of *Talaromyces* spp. and *P. purpurogenum rubisclerotium*, respectively) and their liberation as extrolites in the liquid broth (up to 278 and 1118 meqv·L^{-1} of culture filtrate of *Talaromyces* spp. and *P. purpurogenum rubisclerotium*, respectively).

Table 1. Biomass and extracellular versus intracellular pigment production by the marine isolate *Talaromyces* spp. and terrestrial isolate *P. purpurogenum rubisclerotium* in submerged cultures.

Fungal Strain	Broth	Biomass	Extracellular Pigments	*a*-Value (CIELab)	Intracellular Pigments Extracted	
		(g/L)	(meqv. Carmine/L)	(Red Color)	(meqv/g Biomass)	or (meqv/L Culture)
30570	in DMD	6.2–7.1	18–31	10.8–13.4	23.8–24.7	148–176
	in PDB	5.0–5.5	208–278	63.8–65.8	49.4–69.8	247–384
	in YCD	1.6–2.0	-	5.3–5.6	44.3–48.5	71–97
4464	in DMD	8.4–8.5	1–189	0.7–0.9	57.9–83.2	487–707
	in PDB	1.0–1.3	685–1118	49.9–51.5	69.2–116.0	90–116
	in YCD	1.6–1.7	666–680	42.8–43.5	22.5–71.2	36–121

30570: marine isolate of *Talaromyces* spp.; 4464: terrestrial isolate of *Penicillium purpurogenum rubisclerotium*; DMD: Defined Minimal Medium broth; PDB: Potato Dextrose Broth. YCD: Yeast Casamino acids Dextrose Broth; Values corresponding to the main productivities are noticed in bold.

The intracellular extracts of the *Talaromyces* spp. marine isolate and *P. purpurogenum rubisclerotium* terrestrial isolate were analyzed by HPLC-DAD. Concerning the polyketide pigments extracted from the mycelial biomass of *Talaromyces* spp. grown in PDB culture, all the intracellular extracts recovered from the six-stage PLE presented an intense purple-red shade (S4). The chromatograms of the intracellular extracts shown in Figure 4 indicated that the intracellular aqueous extract (Figure 4C) was entirely composed of a major colored metabolite (compound **1**; not tentatively identified, >95% w/w on total compounds in this intracellular aqueous extract) with the retention time (Rt.) of 1.71 min (red polar compound with λ_{max} 215, 244, 276, 418, 514, 524 nm). Then, the 50% aqueous methanolic solution used as a solvent in the six-stage PLE extracted the largest part of other intracellular pigments from *Talaromyces* spp. Six other major colored metabolites were detected in this reddish liquid sample (Figure 4C): compound **2** with Rt. 26.07 min (3.1% w/w on total secondary metabolites in this intracellular hydroalcoholic extract), compound **3** with Rt. 29.60 min (4.9%), the main compound **4** with Rt. 30.97 min (54.4%), compound **5** with Rt. 32.66 min (14.8%), compound **6** with Rt. 38.04 min (10.9%), and compound **7** with RT 43.95 min (6.3%). In addition, compound **8** with Rt. 69.78 min, identified as ergosterol by UHPLC-HRMS analysis (see below), was essentially detected in the 100% methanolic extract (solvent polarity index of 5.0) (Figure 4E).

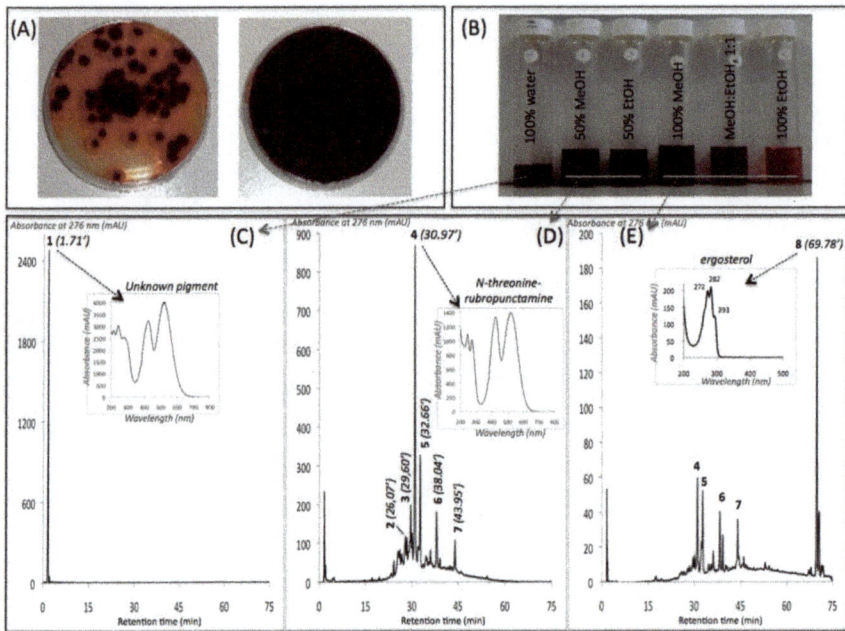

Figure 4. (A) Cultures in PDA of the marine isolate *Talaromyces* spp. 305_70; **(B)** liquid extracts obtained after a six-stage pressurized liquid solvent extraction of the mycelium cultivated in PDB submerged culture; **(C–E)** chromatograms* of the overall compounds detected by HPLC-DAD in the extracts. *Captions*: PDA (Potato Dextrose Agar); PDB (Potato Dextrose Broth); HPLC-DAD: high-performance liquid chromatogram combined with photo-diode array detection; Rt.: retention time; * the analysed samples were extracted with different solvents. Table 2 reports on all of the identified or assumed compounds detected in all the different samples and, in particular, the 50% aqueous methanolic extract was the most representative, as shown in Figure 4D. Only compound 1 (Rt. 1.71 min), not tentatively identified, was detected in the intracellular aqueous extract (Figure 4C). The main compound 4 (Rt. 30.97 min), identified as *N*-threoninerubropunctamine, was detected mainly in the 50% aqueous methanolic extract (Figure 4D), and compound 8 (Rt. 69.78 min), identified as ergosterol, was detected mainly in 100% methanolic extract (Figure 4E).

The eluted compounds (from compound **1** to **8**) were analyzed by performing UHPLC-HRMS, their absorption spectra were measured from 200–800 nm, and their mass spectra were determined and compared against the spectral library and literature database for identification. The UV-visible absorption and mass spectra of each identified or assumed compound detected in all the different intracellular extracts of the marine isolate *Talaromyces* spp. are shown in Figure 5. The results suggested that the main compound **4** that presents two absorption maxima in the visible region, at λ_{max} at 424 and 521 nm (red color), was identified as the red polyketide azaphilone pigment *N*-threoninerubropunctamine. Indeed, the molecule showed a similar UV-visible absorption spectrum, as well as an ESI-MS molecular ion in positive mode $[M + H]^+$ at m/z 456 as the red pigment *N*-threoninerubropunctamine previously isolated from some species of *Talaromyces* by other authors [25,26]. Presumably, the compound **3** that presents λ_{max} at 421 and 518 nm in the visible region, and the ESI-MS molecular ion observed at m/z 416 in positive mode $[M + H]^+$ seemed to correspond to the red polyketide azaphilone pigment glycylrubropunctatin, which has previously been isolated from *Monascus* cultures [27]. Compound **6** (*N*-glutarylrubropunctamine) and **7** (monascorubramine), which exhibit ESI-MS molecular ions observed at m/z 484 and m/z 381 in positive mode $[M + H]^+$, respectively, were identified in good agreement with the expected mass of the corresponding molecules previously isolated from some species of *Talaromyces* [25,26]. Finally, the results suggested that compound **8** (with Rt. 69.78 min) seemed to correspond to an ergosterol derivate, according to its absorption spectrum, which was similar to that of an ergosterol molecule, and to the ESI-MS molecular ion observed at m/z 393 in positive mode $[M + H]^+$ relatively close to those of the standard molecule (which in addition to the expected ion m/z 397, also yielded the same ion at m/z 393 (major) because ergosterol had undergone desaturation during LC/MS) [28]. This observation is consistent with the findings of previous studies, which have already isolated ergosterol derivates from some fungal species [29,30]. All secondary metabolites detected from intracellular extracts of *Talaromyces* spp. are reported in Table 2, and the chemical structures of all identified or assumed compounds are described in Figure 6.

The three other compounds eluted, i.e., compound **1** (with Rt. 1.71 min; λ_{max} 215, 244, 276, 418, 514, 524 nm; but no signal observed in ESI-MS analysis), compound **2** (with Rt. 26.07 min; λ_{max} 246, 276, 425, 512 nm; ESI-MS molecular ion observed at m/z 488 in positive mode $[M + H]^+$ together with the ion at m/z 550 $[M+Na+CH_3CN]^+$), and compound **5** (with Rt. 32.66 min; λ_{max} 218, 250, 287, 581 424, 546 nm; ESI-MS molecular ion observed at m/z 498 in positive mode $[M + H]^+$ together with the ion at m/z 546), presented UV-visible absorption spectra relatively close to those of the *Monascus*-like azaphilone red pigments, which exhibit two absorption maxima in the visible region near 420 and 520 nm (Figure S5). According to this information, these three not tentatively identified metabolites were classified as water-soluble red azaphilone-derivative pigments. Unfortunately, further experiments with a dereplication purpose using UHPLC-HRMS and comparison with a spectral library were not conclusive enough to fully conclude on the nature of these pigmented compounds. Works are in progress in our laboratory to fully characterize the pigments by NMR analysis.

Table 2. Overall compounds detected by HPLC-DAD and UHPLC-HRMS in intracellular extracts (IE) of the marine isolate of *Talaromyces* spp. and terrestrial isolate of *P. purpurogenum rubisclerotium* cultivated in PDB submerged culture, with reference to the chromatograms shown in Figures 4 and 7.

Compound No.	R.t. (mn)	λmax (nm)	Solvent	Tentative Identification	Mol. Peak (m/z)
\multicolumn{6}{c}{IE of *Talaromyces* spp. (305_70)}					
1	1.71	215, 244, 276, 418, 514, 524	water	n.i.	n.d.
2	26.07	246, 276, 425, 512	MeOH	n.i.	488 $[M + H]^+$
3	29.60	245, 274, 421, 518	MeOH	Glycylrubropunctatin	416 $[M + H]^+$
4	**30.97**	**245, 274, 424, 521**	**MeOH**	**N-threoninerubropunctamine**	**456 $[M + H]^+$**
5	32.66	218, 250, 287, 424, 546	MeOH	n.i.	498 $[M + H]^+$
6	38.04	246, 273, 427, 521	MeOH	N-glutarylrubropunctamine	484 $[M + H]^+$
7	43.95	245, 272, 424, 519	MeOH	Monascorubramine	381 $[M + H]^+$
8	69.78	272, 282, 293	MeOH	Ergosterol	393 $[M + H]^+$
\multicolumn{6}{c}{IE of *Penicillium purpurogenum rubisclerotium* (4464)}					
8	69.78	272, 282, 293	MeOH	Ergosterol	393 $[M + H]^+$
9	1.57	216, 292, 492	MeOH	n.i.	n.d.
10	**22.05**	**262, 322, 430**	**MeOH**	n.i.	n.d.
11	23.50	284, 389	MeOH	n.i.	n.d.

IE: Intracellular extract; PDB: Potato Dextrose Broth; Main compound in bold; n.i. = not tentatively identified.

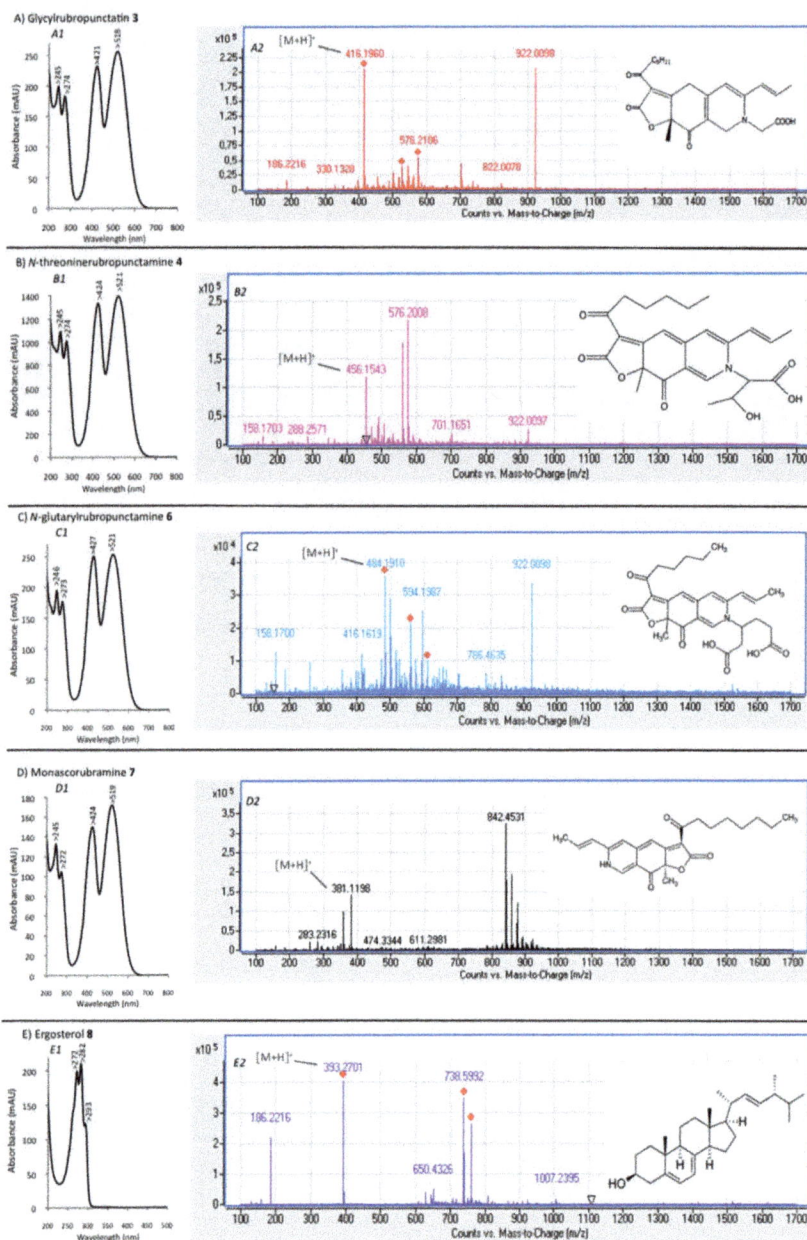

Figure 5. UV-visible absorption (**A1**, **B1**, **C1**, **D1**, **E1**) and mass spectra (**A2**, **B2**, **C2**, **D2**, **E2**) of the identified or assumed compounds detected in intracellular extracts of the marine isolate *Talaromyces* spp. with reference to the chromatograms shown in Figure 4: glycylrubropunctatin **3**, *N*-threoninerubropunctamine **4**, *N*-glutarylrubropunctamine **6**, Monascorubramine **7**, and ergosterol **8**. Compounds **1**, **2** and **5** are red azaphilone-derivative pigments not tentatively identified.

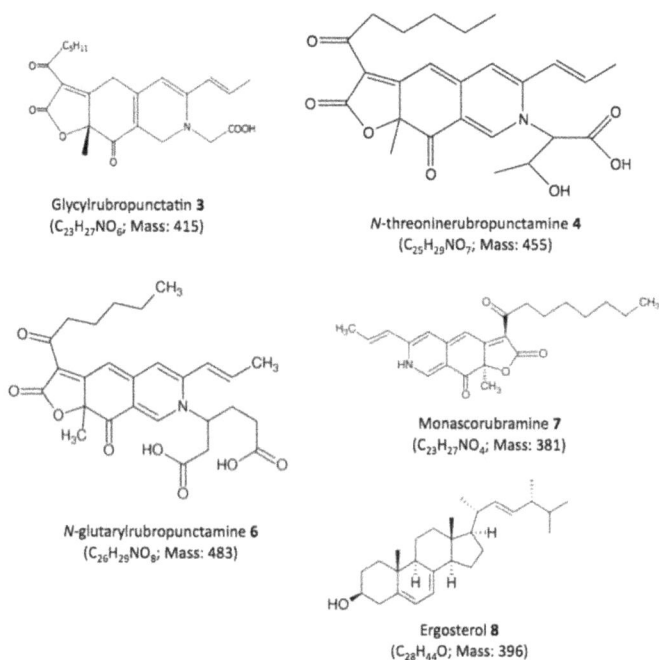

Figure 6. Chemical structures of the identified or assumed compounds detected in the present study in intracellular extracts of the marine isolate *Talaromyces* spp. Formula and calculated nominal masses are shown in parentheses.

For the purpose of comparison, Table 2 summarized the compounds eluted in all the different intracellular extracts of the strain *P. purpurogenum rubisclerotium*. Regarding the intracellular pigments extracted from this fungal mycelium cultivated in PDB, the chromatograms shown in Figure 7 indicated that the aqueous extract was entirely composed of a major secondary metabolite with Rt. 1.57 min (compound **9** with the following λ_{max}: 216, 292, 492 nm; not tentatively identified), which is different from compound **1** detected from the aqueous extract of *Talaromyces* spp. Similarly, the 50% aqueous methanolic solution used as a solvent in the six-step PLE extracted the largest part of other intracellular pigments from this *P. purpurogenum rubisclerotium* strain: two other major colored metabolites were detected in this reddish liquid extract: the main compound **10** with Rt. 22.05 min that presents λ_{max} at 262, 322, 430 nm (not tentatively identified); and compound **11** with Rt. 23.50 min exhibited λ_{max} at 284 and 389 nm (not tentatively identified). Further analytical experiments and comparisons with the spectral library were not conclusive enough to fully conclude on the nature of these pigmented compounds from intracellular extracts of *P. purpurogenum rubisclerotium*. The last compound eluted with Rt. 69.78 min was essentially detected in the 100% methanolic extract, and seemed to correspond to the same ergosterol **8** derivate previously isolated from intracellular extracts of *Talaromyces* spp. marine isolate.

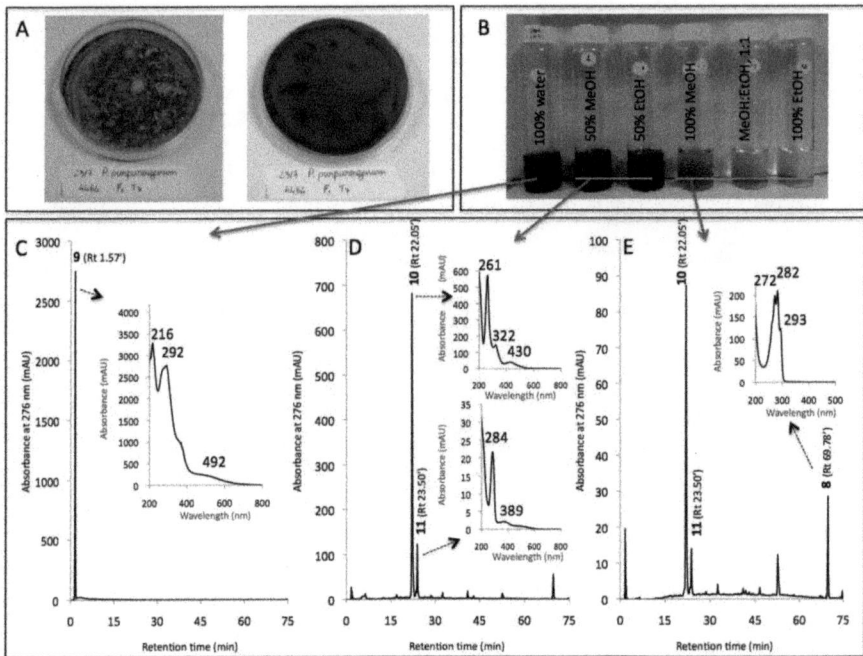

Figure 7. (**A**) Cultures in PDA of the strain *Penicillium purpurogenum rubisclerotium* LCP4464; (**B**) liquid extracts obtained after a six-stage pressurized liquid solvent extraction of the mycelium cultivated in PDB submerged culture; (**C–E**) chromatograms * of the overall compounds detected by HPLC-DAD in the different extracts. *Captions:* PDA (Potato Dextrose Agar); PDB (Potato Dextrose Broth); HPLC-DAD: high-performance liquid chromatography combined with photo-diode array detection; Rt.: retention time; *the analysed samples were extracted with different solvents. Table 2 reports the compounds detected in all the different samples and, in particular, the 50% aqueous methanolic extract was the most representative, as shown in Figure 7D. Only compound **9** (Rt. 1.57 min; not tentatively identified) was detected in the aqueous extract (Figure 7C), the main compound **10** (Rt. 22.05 min; not tentatively identified) was detected mainly in the 50% aqueous methanolic extract (Figure 7D), and ergosterol 8 (Rt. 69.78 min) was detected mainly in 100% MeOH extract (Figure 7E).

4. Discussion

The 11 fungal strains from the terrestrial environment investigated in this study were firstly selected according to the capacity of analogous fungal species cited in the literature to produce pigments in solid and/or submerged cultures. The four local marine isolates were selected according to previous works performed in our laboratory (data not published) based on their extracellular pigment producing ability in submerged cultures. We identified four strains, *P. purpurogenum rubisclerotium*, *F. oxysporum*, and two marine isolates sampled from the lagoon on the west coast of La Reunion Island identified as *Talaromyces* spp. and *Trichoderma atroviride*, as potential pigment producers that produced polyketide red pigments. Regarding the pigment yield, there are generally two natural ways in which the concentration of pigments can be increased: either by improving the fungal growth, or by increasing the intracellular accumulation of pigments. The problem, however, lies with the difficulty in increasing both the biomass and the pigment production, which would be optimal for industrial production. Biomass and pigment yields tend to be negatively correlated. An increase in the biomass yield is connected to the abundance of nutrients in the medium, whereas pigment production tends to be increased under nutrient-poor conditions and/or external stresses, as a protective mechanism [31].

Therefore, the relationship between biomass and pigment production needs to be fully elucidated in order to optimize and control pigment production in submerged fungal culture.

Our results confirmed that same fungal strain did not show the same metabolism of extracellular pigments in submerged cultures (due to the excretion of diffusible pigments in liquid broth) with respect to the medium composition used. For example, the highest water-soluble extrolites production was obtained for the culture filtrate of the strain *P. purpurogenum rubisclerotium* in PDB liquid broth, although its biomass concentration was very low in this medium. As a result, the nutrient composition of the PBD liquid broth allows one to associate the high production and excretion of extracellular water soluble pigments with the low formation of mycelial biomass. Along similar lines, two other strains belonging to the reddish pigment producers that are *F. oxysporum* and *T. atroviride* presented the highest extracellular polyketide pigment productivities in PDB submerged cultures, but the lowest contents of mycelial biomass. These results confirmed the difficulty in increasing both the mycelial biomass and the extracellular pigment production at the same time, which would be optimal for industrial production. Surprisingly, the marine isolate *Talaromyces* spp. presented a different behavior towards the medium nutrient composition. It simultaneously showed a relatively high mycelial biomass content (above 5.5 g·L^{-1} of dry biomass) and a strong content in extracellular polyketide pigments in culture filtrate, using submerged fermentation in PDB.

In general, the secretion of extracellular pigments is favored over intracellular production at an industrial scale, as it implies less downstream extraction processes. Furthermore, there are several submerged fermentation techniques that can be used, such as fed-batch or continuous mode approaches, in order to achieve optimal pigment yields [31]. Our results suggested that an increase in the mycelial biomass yield is mainly connected to the abundance of simple source of carbon and nitrogen in the medium (f.i. in DMD broth), whereas the production and especially secretion of extracellular polyketide pigments seemed to be increased with an abundance of complex sources of carbon (e.g., starch from PDB) and nitrogen (e.g., amino acids and proteins from PDB) combined with some essential cofactors like calcium, magnesium, phosphor, iron, manganese, zinc, and copper (all present in PDB broth). Indeed, it has been previously reported in the literature that the use of a complex nitrogen source or the addition of individual amino acids tends to influence the number, type, and excretion rates of different pigments that are formed in submerged fungal cultures [31]. It has been suggested that the stimulating effect of amino acids on the production or excretion of pigments is caused by an increase in solubility due to the greater hydrosolubility profile of the pigments when bound to an amino acid than the native pigment on its own. It has also been reported that fungal pigments access the aqueous environment by association with proteins or other polar compounds [31].

In this study, the volumetric production of polyketide extrolites secreted by each strain in the culture filtrate was estimated by a spectrophotometer on the basis of the measured absorbance at 276 nm, and expressed as milli-equivalents of polyketide carmine per liter of liquid broth. This estimation is a value proportional to the total polyketide pigment concentration in the culture filtrate. Indeed, most polyketide-derived pigments are characterized by absorption bands in the UV domain (near 240–260 nm) due to the benzene structure and most of them also presented one UV absorption peak near 276 nm, while the maximum absorbencies in the visible region of these polyketide-derived pigments varied from the range of 400 (yellow pigments) to 500 nm (red pigments). Examples include the anthraquinone red pigment carmine which exhibited a λ_{max} at 276 nm, the naphtoquinone red pigment bikaverin (λ_{max} at 253 and 276 nm) isolated from *Fusarium* species, and most *Monascus*-like pigments with a common azaphilone skeleton isolated from *Penicillium* and *Talaromyces* species such as yellow ankaflavin and red monascorubramine (λ_{max} at 276 nm). Furthermore, our data indicate that the positive a*-value, i.e., the red colorimetric coordinate in the CIE L*a*b* color system, and the measured absorbance at 276 nm of the pigmented culture filtrates, tend to be slightly positively correlated for some fungal strains belonging to the reddish pigment producers (data not shown), which justifies the use of this wavelength as the detection value for all polyketide pigments produced by the fungal strains.

For the quantification and characterization of intracellular polyketide pigments, a greener protocol for the pigment extraction, that is a pressurized liquid extraction (PLE) procedure using environmental-friendly solvents, has been investigated. To our knowledge, there is no standard method for polyketide-derived pigment extraction from mycelial biomass. However, based on literature works, the commonly used process involves mixing dried biomass with an organic solvent (like ethyl acetate, chloroform, etc.) or a mixture, followed by the mechanical disruption of the cells, and subsequent centrifugation or filtration (solid-liquid extraction) [6,31]. In general, cell disruption is most necessary for the higher recovery of intracellular fungal pigments. Many different methods for cell disruption have been suggested by authors, including mechanical (sonification) and non-mechanical disruptions (chemical extraction processes) [6,31]. The preferred extraction procedure is based on a quick process that efficiently releases all of the intracellular pigments from the matrix into the solution without altering them, while using environmental-friendly solvents, if possible [6,31]. Moreover, the polarity of the biocompounds to be extracted is the determinant in the selection of the extraction solvent [32]. Therefore, and as the fungal biomass is likely to contain a mixture of pigments of various natures, multi-stage extraction methods using solvents of a different polarity are required, in order to obtain the most exhaustive pigment composition profile for each fungal strain tested.

In our study, lyophilized mycelial biomass was subjected to a six-stage PLE using environmental-friendly solvents including water, methanol, and ethanol, all already allowed and widely used in the EU and in the US for the extraction of natural food colorants. The protocol was illustrated in the intracellular pigment extracts of *P. purpurogenum rubisclerotium* and *Talaromyces* spp. analysed by HPLC-DAD and UHPLC-HRMS. The PLE system offers the advantages of considerably reducing the extraction times and solvent amounts to be used, as well as using more sustainable solvents such as water, ethanol, and methanol with higher extraction efficiencies [32]. It consists of a solid-liquid extraction process carried out at a high temperature (50–100 °C) and elevated pressure (10–150 MPa) in order to maintain the solvent(s) at a liquid state when applied to the sample [18,19]. The high pressures help in forcing the solvent into the cell pores, while the increased temperature enhances the extraction kinetics. Additionally, the use of increased pressure during the extraction protects the compounds of interest from oxygen, and ensures a higher quality of the recovered biochemical. Then, PLE is considered as a promising extraction process particularly appropriate for polar compounds. Moreover, being able to use common and non-toxic solvents is of great interest when dealing with compounds, which are intended to be used in food and cosmetic products, from health and safety, environmental, and economic prospects, as it would result in less downstream purification processes, less waste treatments, and less energy and solvent expenses. Furthermore, solvents such as ethanol and methanol can be produced from carbon neutral homoacetogenic gas fermentation and biogas produced from waste, respectively, strengthening the sustainability of such process. The six-stage PLE procedure gives encouraging results in terms of the efficiency and selectivity of the pigments extracted from the fungal mycelia. It also paves the way for further optimizations of the solvent mixture which can be used to isolate specific pigments. In the meantime, extraction results confirmed that the minimal 'DMD' medium favors the production of mycelial biomass to the detriment of pigment production. On the other hand, the nutrient composition of the PDB medium favors both the intracellular production of pigments (up to 69.8 and 116.0 meqv·g^{-1} dry biomass of *Talaromyces* spp. and *P. purpurogenum rubisclerotium*, respectively) and their liberation as extrolites in the liquid broth (up to 277 and 1178 meqv·L^{-1} culture filtrate in submerged culture of *Talaromyces* spp. and *P. purpurogenum rubisclerotium*, respectively). Thus, an efficient and sustainable multi-step extraction process has been developed, allowing a sequential extraction of a wide panel of pigmented molecules based on their respective polarity. Such a process could be further applied to other types of sample matrices for biochemical content determination.

The 50% aqueous methanolic solution used as extraction solvent in the six-stage PLE procedure recovered the largest part of intracellular pigments from the two fungal mycelia. Seven major colored metabolites were detected in intracellular pigment extracts of *Talaromyces* spp. marine isolate: all corresponding to *Monascus*-like pigments with a common azaphilone skeleton, including the main water-soluble azaphilone red pigment identified as *N*-threoninerubropunctamine **4** (non-toxic compound), followed by three other assumed *Monascus*-like azaphilone pigments such as *N*-glutarylrubropunctamine **6**, monascorubramine **7**, and glycylrubropunctatin **3**. The generally recognized pathway proposes that the orange *Monascus* and *Monascus*-like azaphilone pigments, including monascorubrin and rubropunctatin (insoluble in water), are formed by the esterification of a beta-ketoacid (from the fatty acid synthase pathway) to the chromophore (derived from the polyketide synthase pathway). Then, the reduction of the orange *Monascus* and *Monascus*-like azaphilone pigments yields the yellow *Monascus*-like azaphilone pigments (monascin, ankaflavin). In contrast, amination of the orange pigments with amino group-containing compounds in the medium (proteins, amino acids and nucleic acids) leads to the water-soluble red *Monascus* and *Monascus*-like azaphilone pigments, including the monascorubramine **7** and rubropunctamine derivates like *N*-threoninerubropunctamine **4** and *N*-glutarylrubropunctamine **6** identified in this study [33].

For the purpose of comparison, three other red colored metabolites (not tentatively identified in this study) were detected in intracellular extracts of *P. purpurogenum rubisclerotium* terrestrial isolate (also known under its *Talaromyces pinophilus* name, according to Frisvad et al. [14]). Ergosterol **8** was also isolated as a minor compound in the intracellular pigment extracts from the two fungal mycelia.

5. Conclusions

We report on this article newly isolated polyketide red pigment producing ascomycetous fungi namely *Penicillium purpurogenum rubisclerotium*, *Fusarium oxysporum*, and two fungal isolates of marine origin identified as *Talaromyces* spp. and *Trichoderma atroviride*, sampled from the Reunion Island marine habitat. A new sustainable pigment extraction method was used, that is a six-stage pressurized liquid extraction protocol, for advanced mycelial pigment extraction. Such methods, which use less and non-toxic solvents, have not been used on fungal matrixes thus far. The more promising strain was the non-mycotoxigenic fungus *Talaromyces* spp. of marine origin. According to its pigment production capacity and the emerging natural red pigments in the global market, this fungal strain represents an interesting example of a microorganism which can produce a variety of interesting bioactive compounds like natural red colorants. In addition, the production of secondary metabolites is important for the chemotaxonomic characterization of this fungal isolate. The main pigment produced by this strain has been characterized as *N*-threoninerubropunctamine, a non-toxic red *Monascus*-like azaphilone pigment. Knowledge of the intracellular and extracellular pigment productions by this fungus in submerged culture is important for the development of a high-performance industrial fermentation process for polyketide red pigment production. Further studies are necessary for future applications in the dye industry.

Supplementary Materials: The following are available online at www.mdpi.com/2309-608X/3/3/34/s1.

Acknowledgments: We would like to thank the Conseil Régional de La Réunion, Reunion Island, France, for the financial support of research activities dedicated to microbial pigments. The authors thank Kristian Fog Nielsen of the Department of Systems Biology, Technical University of Denmark, for the UHPLC-HRMS.

Author Contributions: Yanis Caro, Thomas Petit, Mireille Fouillaud, Francesco Vinale and Laurent Dufossé conceived and designed the experiments; Juliana Lebeau and Mekala Venkatachalam performed the experiments. Juliana Lebeau and Yanis Caro analyzed the data and wrote the paper. All authors read and approved the final manuscript.

Conflicts of Interest: The authors declare no conflict of interest.

References

1. Sutthiwong, N.; Caro, Y.; Laurent, P.; Fouillaud, M.; Valla, A.; Dufossé, L. Production of Biocolors. In *Biotechnology in Agriculture and Food Processing: Opportunities and Challenges*, 1st ed.; Panesar, P.S., Marwaha, S.S., Eds.; Francis & Taylor, CRC Press: Boca Raton, FL, USA, 2013; pp. 417–445.

2.   Caro, Y.; Anamale, L.; Fouillaud, M.; Laurent, P.; Petit, T.; Dufossé, L. Natural hydroxyanthraquinoid pigments as potent food grade colorants: An overview. *Nat. Prod. Bioprospect.* **2012**, *2*, 174–193. [CrossRef]
3.   Dufossé, L. Anthraquinones, the Dr Jekyll and Mr Hyde of the food pigment family. *Food Res. Int.* **2014**, *65*, 132–136. [CrossRef]
4.   Durán, N.; Teixera, M.F.; Conti, R.D.; Esposito, E. Ecological-Friendly Pigments From Fungi. *Crit. Rev. Food Sci. Nutr.* **2002**, *42*, 53–66. [CrossRef] [PubMed]
5.   Dufosse, L.; Fouillaud, M.; Caro, Y.; Mapari, S.A.; Sutthiwong, N. Filamentous fungi are large-scale producers of pigments and colorants for the food industry. *Curr. Opin. Biotechnol.* **2014**, *26*, 56–61. [CrossRef] [PubMed]
6.   Caro, Y.; Venkatachalam, M.; Lebeau, J.; Fouillaud, M.; Dufossé, L. Pigments and Colorants from Filamentous Fungi. In *Fungal Metabolites*, 1st ed.; Mérillon, J.-M., Ramawat, K.G., Eds.; Springer International Publishing: Berlin, Germany, 2015; pp. 499–568.
7.   Fouillaud, M.; Venkatachalam, M.; Girard-Valenciennes, E.; Caro, Y.; Dufossé, L. Anthraquinones and Derivatives from Marine-Derived Fungi: Structural Diversity and Selected Biological Activities. *Mar. Drugs* **2016**, *14*, 64. [CrossRef] [PubMed]
8.   Yilmaz, N.; Houbraken, J.; Hoekstra, E.S.; Frisvad, J.C.; Visagie, C.M.; Samson, R.A. Delimitation and characterisation of *Talaromyces purpurogenus* and related species. *Persoonia* **2012**, *29*, 39–54. [CrossRef] [PubMed]
9.   Santos-Ebinuma, V.C.; Teixeira, M.F.; Pessoa, A., Jr. Submerged culture conditions for the production of alternative natural colorants by a new isolated *Penicillium purpurogenum* DPUA 1275. *J. Microbiol. Biotechnol.* **2013**, *23*, 802–810. [CrossRef] [PubMed]
10.   Mapari, S.A.; Meyer, A.S.; Thrane, U.; Frisvad, J.C. Identification of potentially safe promising fungal cell factories for the production of polyketide natural food colorants using chemotaxonomic rationale. *Microb. Cell. Fact.* **2009**, *8*, 24. [CrossRef] [PubMed]
11.   Mapari, S.A.; Thrane, U.; Meyer, A.S. Fungal polyketide azaphilone pigments as future natural food colorants? *Trends Biotechnol.* **2010**, *28*, 300–307. [CrossRef] [PubMed]
12.   Mapari, S.A.S.; Meyer, A.S.; Frisvad, J.C.; Thrane, U. Production of *Monascus*-like azaphilone pigments. European patent EP2262862 B1, 28 March 2012.
13.   Arai, T.; Koganei, K.; Umemura, S.; Kojima, R.; Kato, J.; Kasumi, T.; Ogihara, J. Importance of the ammonia assimilation by *Penicillium purpurogenum* in amino derivative *Monascus* pigment, PP-V, production. *AMB Express* **2013**, *3*, 19. [CrossRef] [PubMed]
14.   Frisvad, J.C.; Yilmaz, N.; Thrane, U.; Rasmussen, K.B.; Houbraken, J.; Samson, R.A. Talaromyces atroroseus, a new species efficiently producing industrially relevant red pigments. *PLoS ONE* **2013**, *8*, e84102. [CrossRef] [PubMed]
15.   Frandsen, R.J.; Nielsen, N.J.; Maolanon, N.; Sorensen, J.C.; Olsson, S.; Nielsen, J.; Giese, H. The biosynthetic pathway for aurofusarin in *Fusarium graminearum* reveals a close link between the naphthoquinones and naphthopyrones. *Mol. Microbiol.* **2006**, *61*, 1069–1080. [CrossRef] [PubMed]
16.   Limon, M.C.; Rodriguez-Ortiz, R.; Avalos, J. Bikaverin production and applications. *Appl. Microbiol. Biotechnol.* **2010**, *87*, 21–29. [CrossRef] [PubMed]
17.   Gessler, N.N.; Egorova, A.S.; Belozerskaya, T.A. Fungal anthraquinones. *Appl. Biochem. Microbiol.* **2013**, *49*, 85–99. [CrossRef]
18.   Vazquez-Roig, P.; Picó, Y. Pressurized liquid extraction of organic contaminants in environmental and food samples. *Trend. Anal. Chem.* **2015**, *71*, 55–64. [CrossRef]
19.   Carabias-Martínez, R.; Rodríguez-Gonzalo, E.; Revilla-Ruiz, P.; Hernández-Méndez, J. Pressurized liquid extraction in the analysis of food and biological samples. *J. Chromatogr. A* **2005**, *1089*, 1–17. [CrossRef] [PubMed]
20.   Velmurugan, P.; Lee, Y.H.; Venil, C.K.; Lakshmanaperumalsamy, P.; Chae, J.-C.; Oh, B.-T. Effect of light on growth, intracellular and extracellular pigment production by five pigment-producing filamentous fungi in synthetic medium. *J. Biosci. Bioeng.* **2010**, *109*, 346–350. [CrossRef] [PubMed]
21.   Guyomarc'h, F.; Binet, A.; Dufossé, L. Production of carotenoids by *Brevibacterium linens*: variation among strains, kinetic aspects and HPLC profiles. *J. Ind. Microbiol. Biotechnol.* **2000**, *24*, 64–70. [CrossRef]
22.   Dufossé, L.; Galaup, P.; Carlet, E.; Flamin, C.; Valla, A. Spectrocolorimetry in the CIE L*a*b* color space as useful tool for monitoring the ripening process and the quality of PDO red-smear soft cheeses. *Food Res. Int.* **2005**, *38*, 919–924. [CrossRef]

23. Sutthiwong, N.; Caro, Y.; Milhau, C.; Valla, A.; Fouillaud, M.; Dufossé, L. *Arthrobacter arilaitensis* strains isolated from ripened cheeses: Characterization of their pigmentation using spectrocolorimetry. *Food Res. Int.* **2014**, *65*, 184–192. [CrossRef]

24. Klitgaard, A.; Iversen, A.; Andersen, M.R.; Larsen, T.O.; Frisvad, J.C.; Nielsen, K.F. Aggressive dereplication using UHPLC-DAD-QTOF: Screening extracts for up to 3000 fungal secondary metabolites. *Anal. Bioanal. Chem.* **2014**, *406*, 1933–1943. [CrossRef] [PubMed]

25. Jung, H.; Kim, C.; Kim, K.; Shin, C.S. Color Characteristics of *Monascus* Pigments Derived by Fermentation with Various Amino Acids. *J. Agric. Food Chem.* **2003**, *51*, 1302–1306. [CrossRef] [PubMed]

26. Mapari, S.A.S.; Hansen, M.E.; Meyer, A.S.; Thrane, U. Computerized Screening for Novel Producers of *Monascus*-like Food Pigments in *Penicillium* Species. *J. Agric. Food Chem.* **2008**, *56*, 9981–9989. [CrossRef] [PubMed]

27. Yuliana, A.; Singgih, M.; Julianti, E.; Blanc, P.J. Derivates of azaphilone *Monascus* pigments. *Biocatal. Agric. Biotechnol.* **2017**, *9*, 183–194. [CrossRef]

28. Slominski, A.; Wortsman, J.; Plonka, P.M.; Schallreuter, K.U.; Paus, R.; Tobin, D.J. Hair follicle pigmentation. *J. Investig. Dermatol.* **2005**, *124*, 13–21. [CrossRef] [PubMed]

29. Torres, S.; Cjas, D.; Palfner, G.; Astuya, A.; Aballay, A.; Pérez, C.; Hernandez, V.; Becerra, J. Steroidal composition and cytotoxic activity from fruiting body of Cortinarius xiphidipus. *Nat. Prod. Res.* **2017**, *31*, 473–476. [CrossRef] [PubMed]

30. Dame, Z.T.; Silima, B.; Gryzenhout, M.; van Ree, T. Bioactive compounds from the endophytic fungus *Fusarium proliferatum*. *Nat. Prod. Res.* **2016**, *30*, 1301–1304. [CrossRef] [PubMed]

31. Gmoser, R.; Ferreira, J.A.; Lennartsson, P.R.; Taherzadeh, M.J. Filamentous ascomycetes fungi as a source of natural pigments. *Fungal Biol. Biotechnol.* **2017**, *4*, 4. [CrossRef]

32. Azmir, J.; Zaidul, I.S.M.; Rahman, M.M.; Sharif, K.M.; Mohamed, A.; Sahena, F.; Jahurul, M.H.A.; Ghafoor, K.; Norulaini, N.A.N.; Omar, A.K.M. Techniques for extraction of bioactive compounds from plant materials: A review. *J. Food Eng.* **2013**, *117*, 426–436. [CrossRef]

33. Chen, W.; Chen, R.; Liu, Q.; He, Y.; He, K.; Ding, X.; Kang, L.; Guo, X.; Xie, N.; Zhou, Y.; et al. Orange, red, yellow: Biosynthesis of azaphilone pigments in *Monascus* fungi. *Chem. Sci.* **2017**. [CrossRef]

Journal of
Fungi

MDPI

Article

Biodiversity of Pigmented Fungi Isolated from Marine Environment in La Réunion Island, Indian Ocean: New Resources for Colored Metabolites

Mireille Fouillaud [1,2,*], Mekala Venkatachalam [1], Melissa Llorente [1], Helene Magalon [3], Pascale Cuet [3] and Laurent Dufossé [1,2]

[1]   Laboratoire de Chimie des Substances Naturelles et des Sciences des Aliments—LCSNSA EA 2212, Université de La Réunion, 15 Avenue René Cassin, CS 92003, F-97744 Saint-Denis CEDEX 9, Ile de La Réunion, France; mekalavenkat@gmail.com (M.V.); melissa.llorente@gmail.com (M.L.); laurent.dufosse@univ-reunion.fr (L.D.)

[2]   Ecole Supérieure d'Ingénieurs Réunion Océan Indien—ESIROI, 2 Rue Joseph Wetzell, F-97490 Sainte-Clotilde, Ile de La Réunion, France

[3]   UMR ENTROPIE and LabEx CORAIL, Université de La Réunion, 15 Avenue René Cassin, CS 92003, F-97744 Saint-Denis CEDEX 9, Ile de La Réunion, France; helene.magalon@univ-reunion.fr (H.M.); Pascale.cuet@univ-reunion.fr (P.C.)

*   Correspondence: mireille.fouillaud@univ-reunion.fr; Tel.: +2-62-48-33-62

Received: 31 May 2017; Accepted: 28 June 2017; Published: 2 July 2017

Abstract: Marine ecosystems cover about 70% of the planet surface and are still an underexploited source of useful metabolites. Among microbes, filamentous fungi are captivating organisms used for the production of many chemical classes of secondary metabolites bound to be used in various fields of industrial application. The present study was focused on the collection, isolation, screening and genotyping of pigmented filamentous fungi isolated from tropical marine environments around La Réunion Island, Indian Ocean. About 150 micromycetes were revived and isolated from 14 marine samples (sediments, living corals, coral rubble, sea water and hard substrates) collected in four different locations. Forty-two colored fungal isolates belonging to 16 families, 25 genera and 31 species were further studied depending on their ability to produce pigments and thus subjected to molecular identification. From gene sequence analysis, the most frequently identified colored fungi belong to the widespread *Penicillium*, *Talaromyces* and *Aspergillus* genera in the family Trichocomaceae (11 species), then followed by the family Hypocreaceae (three species). This study demonstrates that marine biotopes in La Réunion Island, Indian Ocean, from coral reefs to underwater slopes of this volcanic island, shelter numerous species of micromycetes, from common or uncommon genera. This unstudied biodiversity comes along with the ability for some fungal marine inhabitants, to produce a range of pigments and hues.

Keywords: fungi; biodiversity; Indian Ocean; Marine; coral reef; genotyping; pigment production

---

1. Introduction

With the growing demand for natural compounds in the industrial sector, marine derived fungi appear to present many interests. Filamentous fungi are ubiquitous in nature due to their huge capacity of adaptation and their ability to produce an assortment of new secondary metabolites. Literature now abundantly reports the significant involvement of fungi in the industry, through the production of various useful substances, such as antibiotics, immunosuppressants, anti-cancer drugs, plant hormones, enzymes, acids and also natural pigments [1–5]. Both the pigments and enzymes equally find their usages in food and beverages, animal feeds, pharmaceuticals, cosmetics, textile, leather, pulp and paper industries, biofuel production, and environment bioremediation [6].

Nevertheless, the distribution of the marine-derived fungal species and their contribution to marine biotopes are still in infancy, and more has to be explored [7–11]. The highest diversity of marine fungi seems to appear in tropical regions, mainly in tropical mangroves, which are extensively studied because of their high richness in organic matters and especially lignocellulosic materials, favorable to the development of a wide range of heterotrophic microorganisms [11–14]. Anyway, many marine ecological niches are still unexplored and it seems plausible that unique features of marine environments can be the inducers of unique substances, biosynthesized by marine or marine-derived microorganisms [15,16].

Considering the immense genetic and biochemical diversity of these fungi, partially derived from the specificity of the biotopes they are facing, marine-derived fungi are regarded as a potential bright source of new molecules with likely application in pigment production [17,18]. Many genera producing pigments have then been isolated either from water, sediments, and decaying organic residues, or from living organisms such as invertebrates, plants or algae. Fungi belonging to genera such as *Aspergillus, Penicillium, Paecilomyces, Eurotium, Alternaria, Fusarium, Halorosellinia, Monodictys* and *Microsphaerospsis* have already been identified from marine biotopes [19–21]. They are therefore able to exhibit bright colors, from yellow to black, mainly belonging to polyketides. Indeed, polyketides pigments and particularly azaphilones and anthraquinones seem to dominate marine natural products of fungal origin [22]. Colored compounds, usually described as secondary metabolites, do not seem to be directly involved in the primary growth of the fungus in which they occur [23]. However they may play some important roles in the resistance to a variety of adverse environmental factors (desiccation, exposure at extreme temperatures, irradiations and photo-oxidation) or in ecological interactions with other organisms (macroorganisms such as sponges, corals or other microbial communities) [24]. For this reason, many fungal secondary metabolites exhibit useful biological activities and are of interest to the pharmaceutical, food, and agrochemical industries [16,25].

This study initiated the search for filamentous fungi in some tropical marine biotopes of coral reefs and underwater slopes of the volcano from La Réunion Island. Fungal isolates from samples of sediments, seawater, hard substrates, coral rubbles or living coral individuals (*Pocillopora* sp.) were characterized both by phenotypic and molecular ways. The production of pigments of quinoid-type produced from the mycelia cultured in liquid media was used as a first approach to screen for the pigment production. This work reveals a part of the mycofloral biodiversity in La Réunion Island tropical marine environment and its potentiality to propose new pigment sources to expand in an industrial setting.

2. Materials and Methods

2.1. Samples Collection

La Réunion Island lies in the Indian Ocean and is located 800 km east of Madagascar (21°06′54.5″ S and 55°32′11.0″ E) (Figure 1a). This tropical island arose two million years ago from a volcanic hot spot (Piton de La Fournaise) and is known for its rainforested interior and its fringing reefs holding most of the marine wealth.

Figure 1. (a) La Réunion island location (Indian Ocean, 21°06′54.5″ S and 55°32′11.0″ E); (b) geolocation of sampling sites around La Réunion Island (West: La Saline; and East: Sainte Rose and Tremblet); and (c) geolocation of the three sampling spots at La Saline fringing reef: Trou d'eau (TDE inner reef and TDE outer slope) and Planch'Alizé (PA) (back arrow represents the main water flow).

A first set of samples was collected on the fringing reef from La Saline, which lies on the dry west coast of the island. It is more than 9 km long and ranges in width from 50 m in its northern part to 600 m in the south [26]. For the purpose of research, samples were collected from three sampling spots on the west coast that cover the sites of Trou d'Eau (TDE) (inner reef flat at −1 m depth and outer slope at −17 m) and Planch' Alizé (PA) (inner reef flat, −1 m) (Figure 1b,c). Planch' Alizé is considered as a sheltered site, located downstream of seawater flowing over the Trou d'Eau (Figure 1c). The outer slope is found at the outer edge of the reef, closest to the open ocean, and is characterized by spurs and grooves extending downward to the sand bottom, while the inner reef flat displays wide transversal strips of branched coral colonies alternating with narrow detrital channels perpendicular to the reef flat [27–31]. Low water flow and high solar radiation contribute to heating the reef water during the day, inducing important daily sea surface temperature variations. This area is also heavily laden with organic and mineral matter coming from nearby human activity (seaside area).

A second set of samples was collected in Sainte Rose area (south-east) on the submerged lava flows (Figure 1b). Indeed, the Piton de la Fournaise is one of the most active effusive volcanoes in the world with 27 eruptions between 1998 and 2007 and a mean frequency, over a century, of an eruptive phase every nine months. Submerged lava flows appear on the south-east part of the island when, during eruptions, the pool of lava overflows the active volcano mouth and pours down on the slopes of the volcano, to the sea. These costal marine ecosystems facing the deep ocean, are then regularly subjected to natural hazards such as being covered by incandescent lava flows, temporary changes in physicochemical conditions of water bodies and exceptional rises of temperature. Besides, this area is poorly inhabited and urbanized and, as a consequence, the amount of organic matter poured in the sea is reduced compared to other coastal ecosystems. It provides a natural laboratory to study the colonization of a blank substrate and the evolution of the biodiversity all around, during the following years. Samples were then obtained from sediments extracted from 1977 lava flow (-25 m depth) and 2004 lava flow (-70 m), as well as from surrounding free water at -70 m.

Seawater, sediments, parts of living corals and hard substrates (volcanic rocks or coral rubbles) were collected in sterile bottles, during the months of April and May 2012, stored in a cooling box (4 °C), brought to the laboratory, and treated immediately for the fungal isolation.

2.2. Culture and Purification of Fungi

To cultivate the revivable fungi from the collected seawater, 100 mL of water was filtered using a 0.45 μm sterile cellulose-nitrate filter (Sartorius Stedim, Göttingen, Germany). The filters were then placed in Petri plates containing malt extract agar (MEA) and Sabouraud agar (BD Difco, Franklin Lakes, NJ, USA) prepared with natural seawater collected near La Saline, and beforehand sterilized at 121 °C, 15 min.

The other samples such as sediments, hard substrates and parts of living/dead coral were treated separately. The samples were first washed with 70% alcohol and rinsed in sterile seawater. Then they were ground using sterile pestle and mortar. Ground material (5 g) was taken from each sample and added to 15 mL of sterile diluent (1.6 g of tryptone (Sigma- Aldrich, T-9410, Saint Louis, MO, USA), 0.05 g of tween 20, 1 L of sterile seawater of pH = 7.5). After stirring for 20 min at 150 rpm on a shaking table (Edmunt Bühler GmbH, VKS 75 Control, Hechningen, Germany), the suspension was diluted by employing serial decimal dilution method up to 10^{-3} [32]. Each diluted sample (1 mL) was poured on Petri plates containing MEA and Sabouraud agar prepared with natural seawater.

All the platings were performed in triplicates and incubated at 25 °C for 21 days. During this period, the plates were checked each day for the appearance of new colonies. Each new colony was individually isolated and cultured on new MEA solid medium. During the growth period, the production of colors was observed.

All the isolated fungi were cultured using monospore technique for future experiments and long-term storage. The fungi grown after 5 days were scraped and transferred into a sterile vial containing a cryoprotectant medium composed of 15% skimmed milk and 2% glycerol for long term storage at -80 °C [33,34]. In total, 42 fungal isolates were then selected for pigment production based on the visual appearance of the thalli grown on solid culture media.

2.3. Fungal Identification

2.3.1. Fungal DNA Extraction

To extract DNA from the 42 purified isolates, a small amount of mycelium along with spores was cultivated on potato dextrose agar (PDA) at 25 °C under day light exposure. After 5 days of growth, the fungal mycelium was scraped and DNA was extracted using DNeasy Blood and Tissue kit (Qiagen, Hilden, Germany) following the manufacturer's instructions. DNA amount and purity contained in each extract were evaluated by measuring the absorbances at 230, 260 and 280 nm (Nanodrop 2000, Thermo Scientific, Waltham, MA, USA) and calculating the ratio A_{260}/A_{280} and A_{260}/A_{230}. DNAs were stored at -20 °C prior to use for amplification studies [35].

2.3.2. Primers, PCR Amplification and Sequencing

The choice of PCR primers was made based on observed phenotypic characteristics for molecular identification. *Aspergillus* species were amplified for calmodulin gene using primers Cmd5/Cmd6 and *Penicillium* species for β-tubulin using primers T10/Bt2b [36]. To amplify and sequence the DNA from *Trichoderma* and *Hypocreales* species, EF-1H/EF-2T primer pair was used to amplify a fragment of the translation elongation factor 1 alpha gene (*Tef1*) [37]. For uncharacterized fungi, the fragments containing ITS region were amplified using ITS1-F_KYO2/ITS2 or ITS3_KYO2/ITS4, and, when necessary, the large subunit rDNA was also amplified using V9G/LR3 primer pair (Table 1) [36–38].

Table 1. PCR amplification and the sequencing primers used for the identification of fungal isolates.

Primers	Direction	Sequences (5′→ 3′)	Note	Hybrid. T °C	Refs.
ITS1-F_KYO2	Forward	TAGAGGAAGTAAAAGTCGTAA		56	
ITS2_KYO2	Reverse	TTYRCTRCGTTCTTCATC		47	[36]
ITS3_KYO2	Forward	GATGAAGAACGYAGYRAA	Small sub-unit, ITS 1, 5.8S, ITS 2, Largest sub unit rDNA	47	
ITS 1	Forward	TCCGTAGGTGAACCTGCGG		55	
ITS 2	Reverse	GCTGCGTTCTTCATCGATGC		55	[39]
ITS 3	Forward	GCATCGATGAAGAACGCAGC		55	
ITS 4	Reverse	TCCTCCGCTTATTGATATGC		55	
V9G	Forward	TTACGTCCCTGCCCTTTGTA	Large sub unit D1/D2 for basidiomycetous yeast	55	
LR3	Reverse	TGACCATTACGCCAGCATCC		57	[38]
Cmd 5	Forward	CCGAGTACAAGGARGCCTTC	Calmodulin, specific for *Aspergillus*	52	
Cmd 6	Reverse	CCGATRGAGGTCATRACGTGG		52	
T 10	Forward	ACGATAGGTTCACCTCCAGAC	β- tubulin, specific for *Penicillium*	55	[38]
Bt2b	Reverse	ACCCTCAGTGTAGTGACCCTTGGC		55	
EF1-728F	Forward	CATCGAGAAGTTCGAGAAGG	Elongation factor 1 for *Trichoderma*	55	[38]
TEF1-LLErev	Reverse	AACTTGCAGGCAATGTGG		55	

PCR reactions were carried out in a total volume of 30 μL: 1× of MasterMix (Applied Biosystems, Foster city, CA, USA), 0.5 μM of forward and reverse primers and at least 1.3 ng/μL of genomic DNA. Amplifications were carried out on a thermal cycler GeneAmp® PCR System 9700 (Applied Biosystems) according to the following program: 94 °C for 5 min + 35 × (94 °C for 30 s, 55 °C (or 52 °C for the primers of calmodulin: Cmd5/Cmd6) for 60 s, 72 °C for 60 s) + 72 °C for 5 min for final elongation step.

2.3.3. Sequence Analysis

Amplicons were sequenced in both directions (GENOSCREEN, Lille, France). The obtained electropherograms were read and corrected with Chromas software (version 2.13, Technelysium pty Ltd., South Brisbane, Australia). The extracted sequences for each gene were separately used to perform nucleotide searches using online BLAST algorithm, provided by NCBI (http://www.ncbi.nlm.nih.gov/BLAST/). BLAST results were sorted based on the maximum identity to the query sequence and considered as the best hit. Sequence-based identities with a cutoff of 97% or above and query coverage >90% were considered as significant [40,41]. Because of low recovery rates and concordance values, some isolates were amplified and sequenced a second time, with additional sets of primers, mainly among the isolates of the genera *Penicillium* and *Trichoderma*.

2.4. Culture Conditions for Pigment Production and Separation of Biomass from Liquid Medium

2.4.1. Culture Conditions

Erlenmeyer flasks (250 mL) containing 80 mL of potato dextrose broth (PDB) medium were autoclaved at 121 °C for 15 min. Then, 120 mg of mycelia from interesting fungal species grown on PDA Petri plates were transferred into the sterile flasks and incubated at 25 °C under daylight exposure, with an agitation of 150 rpm for 10 days (Multitron Pro, Infors HT, Bottmingen, Switzerland).

2.4.2. Separation of Biomass from Culture Liquid

After the end of the fermentation period, the culture medium containing extracellular pigments was separated from mycelia by vacuum filtration using Whatman filter paper No. 2 (Merck, Darmstadt, Germany). Thus, liquid medium and biomass were treated separately. The wet mycelium was further used for the extraction of pigment content.

2.5. Production of Pigments

2.5.1. Determination of Pigments Production in Liquid Cultures

Chromophore is a chemical group that absorbs light of specific frequency and confers color to a molecule. Widespread polyketide pigments such as anthraquinones or azaphilones are often highly substituted aromatic molecules, with fused benzene rings [42]. Thus the majority of the common chromophores from fungi absorb in the UV region (one or several peaks between 200–300 nm), whereas absorbance in the visible region (400–700 nm) highly depends on the nature and the number of substituted groups.

To compare the pigment production of all isolates cultured in PDB medium, the amount of pigments produced in liquids was expressed as mg equivalent (mg eq.) of a chosen commercial standard per liter of culture medium (mg eq. purpurin L^{-1}). Purpurin was chosen as a polyketide pigment in orange-red hue, which absorbs in the UV area (250–270 nm) as many polyketides [43], and also in the visible range 458–520 nm [44]. Thus, the absorbance of an authentic colored standard purpurin (Sigma-Aldrich) was estimated at different concentrations using an UV-visible spectrophotometer (Shimadzu UV-1800 Spectrophotometer). Then, in regard with the diversity of pigments content in the fungal cultures and as a preliminary approach, the absorbance of each sample was measured at 254 nm and the amount of pigments produced was expressed in "mg equivalent purpurin L^{-1}" (Figure S1). In addition, for each isolate, the intracellular (IC) pigments (extracted from the biomass) and extracellular (EC) contents (liquid from culture, separated from the biomass) were scanned between 200 and 600 nm with a UV-1800 spectrophotometer (Shimadzu UV spectrophotometer, Shimadzu Corporation, Kyoto, Japan) in a quartz cell of 10 mm path length.

2.5.2. Extraction of Pigments

IC pigments contained in the wet mycelium were extracted using a methanol: water combination (1:1 v/v) as conventional extraction method. The mixture was immersed in an ultrasonic bath at 45 °C for 30 min. The suspension was allowed to stir overnight at room temperature on a shaking table (VKS 75 Control, Edmunt Bühler GmbH). On the following day, it was filtered through Whatman filter paper No. 2 to recover the solvent containing the pigments extracted from biomass.

To compare the amount of pigments produced within the cells with the one diffused into the extracellular medium, we performed the nonparametric Mann–Whitney–Wilcoxon test as our data did not follow the normal distribution using the R software (R Development Core Team 2016) [45].

3. Results

3.1. Diversity of Isolated Fungi

More than 150 isolates were first recovered from the 14 samples collected among four locations. Among them, 42 were selected for identification, according to their capacity to develop colored mycelia or to secrete colored compounds in the media.

After sequencing, the 42 colored isolates were assigned to 16 families, 25 genera and 31 species (accession numbers mentioned in Table 2). The vast majority of the isolates have been identified with more than 98% concordance rate and recovered with high precision at the species level. However, few fungi (*Acremonium* sp., *Periconia* spp. and *Biscogniauxia* sp.) were identified to the genus level only, according to the gene chosen. The genetic characterization with these primers partially failed for two isolates (*Whalleya microplaca B* and *Wallemia sebi*). *Wallemia sebi* was only characterized according to morphological criteria.

Table 2. Fungal isolates from La Réunion Island marine biotopes from different sample types and sampling sites: Trou d'Eau (TDE); Planch' Alizé (PA); Lava flow corresponds to 1977 lava flow in Sainte Rose/Tremblet area.

Family	Fungal Species	Sampling Site	Gene Accession Number
	Water Bodies		
Davidiellaceae	*Cladosporium Cladosporioides*	Lava flow (−70 m)	JF949719.1
Didymellaceae	*Peyronellaea glomerata* (syn: *Phoma glomerata*)	Lava flow (−70 m)	JQ936163.1
Nectriaceae	*Nectria haematococca* A	Lava flow (−70 m)	XM_003053163.1
Pleosporales Incertae Sedis	*Periconia* sp. A	Lava flow (−70 m)	HQ608027.1
	Periconia sp. B	Lava flow (−70 m)	HQ608027.1
Sporidiobolaceae	*Rhodosporidium paludigenum*	Lava flow (−70 m)	AF444493.1
Stachybotryaceae	*Myrothecium atroviride*	Lava flow (−70 m)	AJ302002.1
Teratospheriaceae	*Hortaea werneckii* (syn: *Cladosporium werneckii*)	Lava flow (−70 m)	JN997372.1
Trichocomaceae	*Aspergillus sydowii* B	Lava flow (−70 m)	KC253961.1
	Emericella qinqixianii	TDE outer slope	AB249008.1
	Penicillium brocae NRRL 32599	TDE outer slope	DQ123642.1
	Penicillium viticola B	TDE inner reef	AB606414.1
	Talaromyces rotundus	TDE inner reef	EU497950.1
	Talaromyces verruculosus	PA inner reef	KC416631.1
Wallemiaceae	*Wallemia sebi*	Lava flow (−70 m)	Morphological Identification
	Living Coral Pocillopora sp.		
Hypocreaceae	*Acremonium* sp.	PA inner reef	FJ770373.1
	Hypocrea koningii	TDE inner reef	JX174420.1
Trichocomaceae	*Aspergillus creber* A	TDE inner reef	JN854049.1
	Aspergillus creber B	TDE inner reef	JN854049.1
	Aspergillus sydowii A	PA inner reef	JN854052.1
	Eurotium amstelodami	TDE outer slope	FR727111.1
	Penicillium viticola C	PA inner reef	AB606414.1
	Coral Rubbles		
Chaetomiaceae	*Chaetomium globosum or Chaetomium murorum*	TDE outer slope	JN209898.1
Trichocomaceae	*Penicillium herquei*	TDE outer slope	JN246042.1
	Talaromyces albobiverticillius B	TDE outer slope	JN899313.1
	Talaromyces albobiverticillius C	TDE outer slope	JN899313.1
	Hard Substrate/Rock Substrate		
Nectriaceae	*Fusarium equiseti* A	Lava flow (−25 m)	JQ936153.1
	Fusarium equiseti B	Lava flow (−25 m)	JF311925.1
	Fusarium equiseti C	Lava flow (−25 m)	JQ936153.1
	Nectria haematococca B	Lava flow (−25 m)	XM_003053163.1
Pleosporaceae	*Epicoccum sorghi* (syn: *Phoma sorghina*; *Peyronellaea stemphylioides*)	Lava flow (−25 m)	KC106717.1
Sordariomycetes	*Nigrospora sphaerica* (or Env. sample from marine air)	TDE outer slope	KC505176.1
Sporidiobolaceae	*Rhodotorula mucilaginosa*	TDE outer slope	KC515367.1
Trichocomaceae	*Penicillium citrinum*	Lava flow (−25 m)	EU030332.1
	Penicillium viticola A	Lava flow (−25 m)	AB606414.1
Ustilaginaceae	*Sporisorium exsertum*	TDE outer slope	JN367293.1
Xylariaceae	*Biscogniauxia* sp.	PA inner reef	FJ884075.1
	Whalleya microplaca A	TDE outer slope	JQ760548.1
	Sordariomycete (or *Whalleya microplaca* B)	TDE outer slope	FJ416301.1
	Sediments		
Didymosphaeriaceae	*Paraconiothyrium variabile*	TDE outer slope	JQ936271.1
Hypocreaceae	*Trichoderma atroviride*	TDE outer slope	KC008065.1
Trichocomaceae	*Talaromyces albobiverticillius* A	TDE outer slope	JN899313.1

Pigment producing fungi (42) were isolated from all types of samples: sediments (3), living coral *Pocillopora* sp. (7), unidentified coral rubbles (4), hard substrates (reef basis or volcanic rocks) (13) and seawater (15) (Table 2).

The most represented fungi, in the selection of colored micromycetes, belonged to the family Trichocomaceae with *Penicillium, Talaromyces* and *Aspergillus* genera (11 species); and then came the Hypocreaceae with *Trichoderma, Hypocrea* and *Acremonium*.

A high diversity of pigmented isolates was observed from the so-called "hard substrates" (rocky basis on which the coral colonies recruit, or submerged lava flows). Some *Nigrospora, Sporisorium, Whalleya*, and *Rhodotorula* isolates were collected from the outer slope at Trou d'Eau (TDE), although they are rather rarely isolated from marine environment. In Sainte Rose, *Penicillium* species (*P. citrinum, P. viticola*) as well as *Fusarium equiseti, Epicoccum sorghi, Nectria haematococca*, were successfully revived from lava flow, sampled at −25 m.

From our study, the coral rubbles (dead parts of corals) contained colored *Penicillium* or related species: *P. herquei* and two isolates of *Talaromyces albobiverticillius* (B and C), as well as an isolate of *Chaetomium globosum*. Coral rubbles or hard substrates naturally appear diversely colored underwater. Indeed, they support the colonization by multiple organisms (colored algae or other aquatic organisms), visually detectable when sampling.

The revivable colored fungi sheltered by the living coral *Pocillopora* sp. belonged to the genera *Aspergillii* (*A. creber, A. sydowii* and *Eurotium amstellodami* (the teleomorphic form of *A. amstellodami*), as well as to *Penicillium* (*P. viticola*), *Hypocrea* (*H. koningii*) and *Acremonium*.

Some fungal species were identified from different types of samples in the same area. At TDE outer slope, *Talaromyces albobiverticillius* A came from sediment and *T. albobiverticillius* B and C were revived from coral rubbles. *Nectria haematococca* was found in lava substrate (−25 m) (isolate B) as well as seawater (−70 m near lava flow) in the same area (isolate A).

Some similar species also appeared in separate locations: *Aspergillus sydowii* was found near lava flow on the east coast (seawater, −70 m) (isolate B) and also in living *Pocillopora* colonies (isolate A), from the west coast back reef (PA site). *Penicillium viticola* was isolated from the west coast on living *Pocillopora* sp. coral in PA (isolate C), from seawater in TDE back reef (isolate B), as well as from lava flow hard substrate (−25 m), on the east coast (isolate A).

These fungi found in several samples and/or in different locations may be considered as frequent in this marine environment.

3.2. Pigment Production

3.2.1. In Culture Broth

The majority of the isolates produced pigments after four days of fermentation in PDB. The colors of the broth (biomass plus liquid culture medium) always darkened over time, which indicated their potential for pigment production (Figure 2).

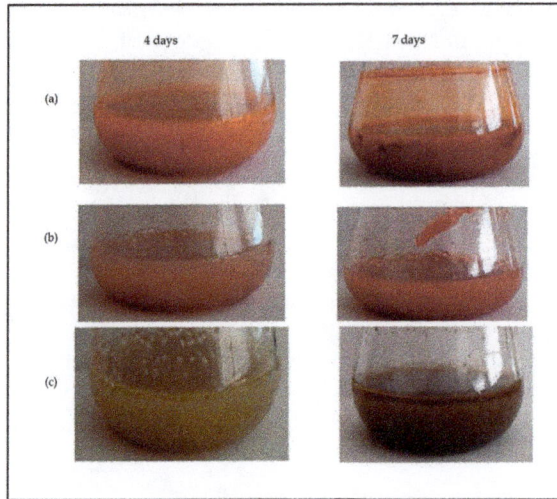

Figure 2. Colors observed in potato dextrose broth cultures from (a) *Talaromyces albobiverticillius* B, (b) *T. albobiverticillius* C, and (c) *Aspergillus creber* B, after four and seven days.

Overall, it was observed that the coloring trend was not directly related to the genus. Even if dominant colors such as yellow, red, brown, purple, orange, pink and green were observed in flasks, the hues were extremely diverse according to the species, even to the isolates (Table 3).

Table 3. Dominant colors of culture broth [1], extracellular (EC) [2] and intracellular (IC) [3] pigments from fungal isolates.

Fungal Isolates	Approximate Hues			Fungal Isolates	Approximate Hues		
Isolates with Intense Hues (Purple/Red/Maroon)				*Isolates with Orange Hues*			
	Broth	EC	IC		Broth	EC	IC
Acremonium sp.				Penicillium viticola A			
Talaromyces albobiverticillius A				Penicillium viticola B			
Talaromyces albobiverticillius B				Epicoccum sorghi			
Talaromyces albobiverticillius C				Penicillium brocae NRRL 32599			
Aspergillus sydowii A				Penicillium herquei			
Aspergillus creber A				Aspergillus sydowii B			
Aspergillus creber B				Chaetomium globosum or C. murorum			
Emericella qinqixianii				Penicillium viticola C			
Trichoderma atroviride				Penicillium citrinum			
Biscogniauxia				Hypocrea koningii			
Paraconiothyrium variabile							
Myrothecium atroviride							
Isolates with Yellow Hues				*Isolates with Green/Brown Hues*			
	Broth	EC	IC		Broth	EC	IC
Peyronellaea glomerata				Talaromyces verruculosus			
Eurotium amstelodami				Talaromyces rotundus			
Rhodosporidium paludigenum				Wallemia sebi			
Periconia sp. A				Sporisorium exertum			
Periconia sp. B				Hortea werneckii			
Rhodotorula mucilaginosa				Whalleya microplaca A			
Fusarium equiseti A				Whalleya microplaca B			
Fusarium equiseti B				Nigrospora sphaerica or Env. sample from marine air			
Fusarium equiseti C				Cladosporium cladosporioïdes			
Nectria haematococca A							
Nectria haematococca B							

[1] Culture broth: mycelium + liquid medium; [2] EC: filtrate from liquid culture medium; [3] IC: intracellular extract of fungal pigments.

Indeed, some similar-looking fungi, identified under a unique accession number (i.e., sharing the same sequence for the considered gene), nevertheless developed different color-phenotypes, while cultured under the same culture conditions. As an example *A. creber* A developed a red hue, clearly different from the green-like color *of A. creber* B ("broth" column in Table 3). Moreover, if the same coloring trend was applicable to all the three isolates of *T. albobiverticillius* (A–C) or *P. viticola* (A–C), different shades of red or yellow-orange hues, respectively, were noticed (Figure 3).

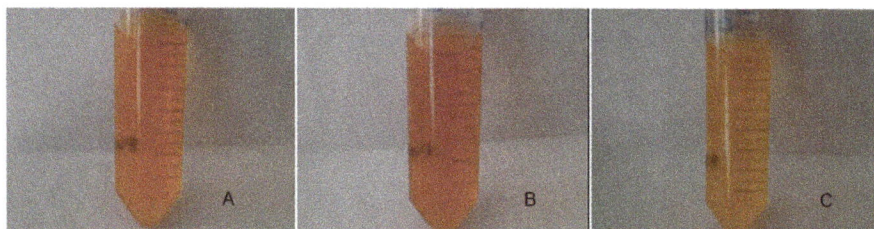

Figure 3. Colors observed from culture filtrates from three isolates *of Penicillium viticola* (**A–C**) (seven-day cultures in potato dextrose broth).

Oppositely, no clear difference could be visually established among the pale pink shades of the three isolates of *F. equisetti* (A–C) or the two *N. haematococca* isolates (A and B).

3.2.2. Pigmented Contents from Mycelium

For all pigment-producing isolates, the intracellular pigments (from mycelium) were extracted from the biomass. The approximate colors visualized after extraction are presented in Table 3. The pigments from most of the extracellular fungal culture filtrates were of dominating red, orange, yellow, green, brown, pink and violet. However, after extraction from biomass, many intracellular samples were uncolored, especially for isolates producing extracellular culture filtrates of pink, yellow and green color. This is probably characteristic of isolates essentially secreting water-soluble colored molecules in the culture media.

Instead, many dark colored cultures, mainly in the shades of red or maroon extracellular pigments, gave dark pigmented intracellular extracts from the biomass, indicating that the pigment was also highly concentrated inside the mycelium. These mainly concerned the isolates included in the group "isolates with intense hues", and in the group "isolates with orange hues" to a lesser extent (Table 3). Thus, isolates appeared with different status and varying capacities, towards pigment production.

3.3. Spectrophotometric Characterization of Pigments

As shown in Figure 4, the absorbance spectra of intra- and extracellular solutions from a single isolate revealed quite similar profiles characterized by a strong absorbance in the UV region and also an area of absorbance in the visible range of wavelengths. The values were principally located in the 400–480 nm area for pale yellow to yellow-orange pigments. The maximal absorbances spread in the 500–550 nm region for red colors.

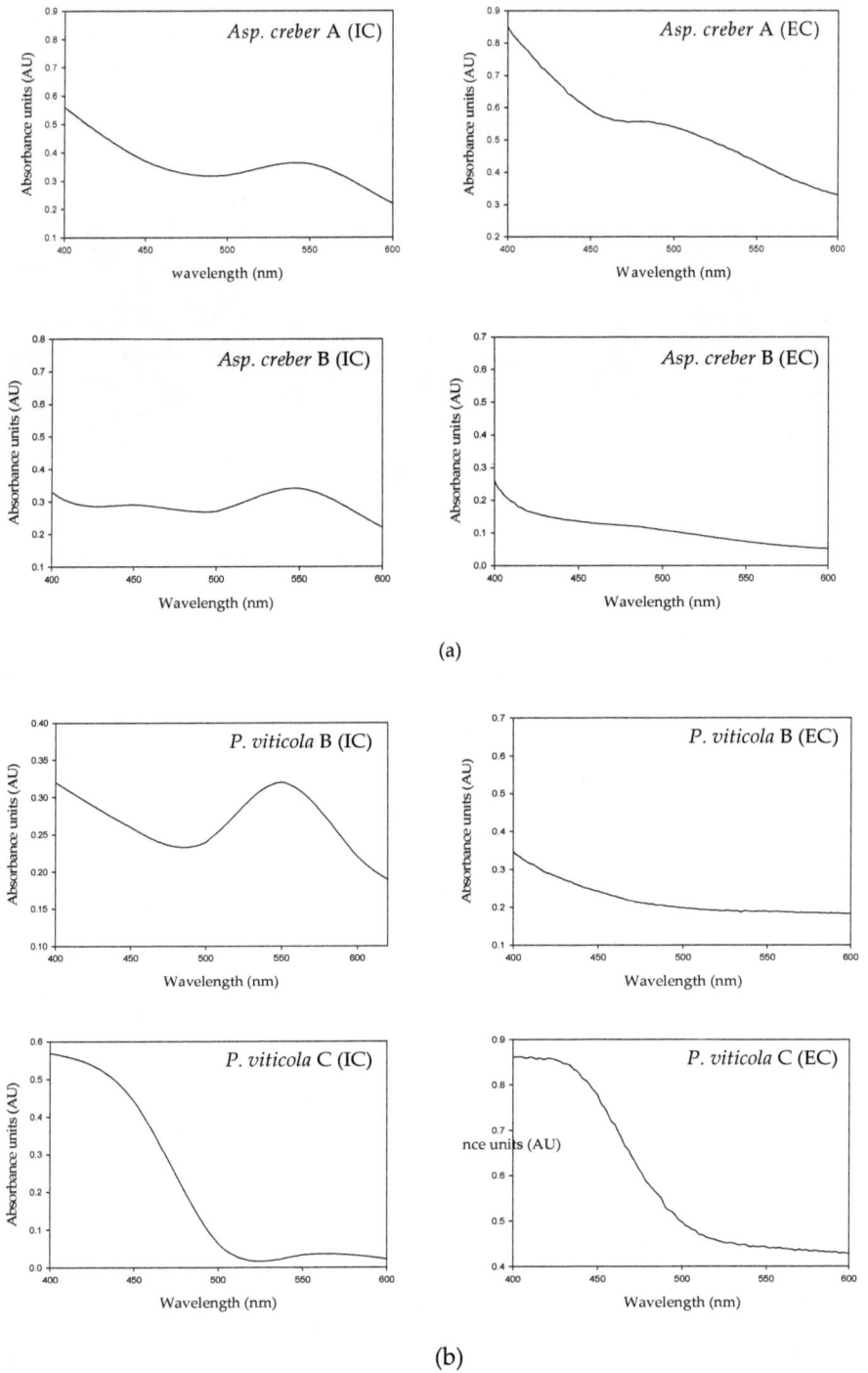

(a)

(b)

Figure 4. *Cont.*

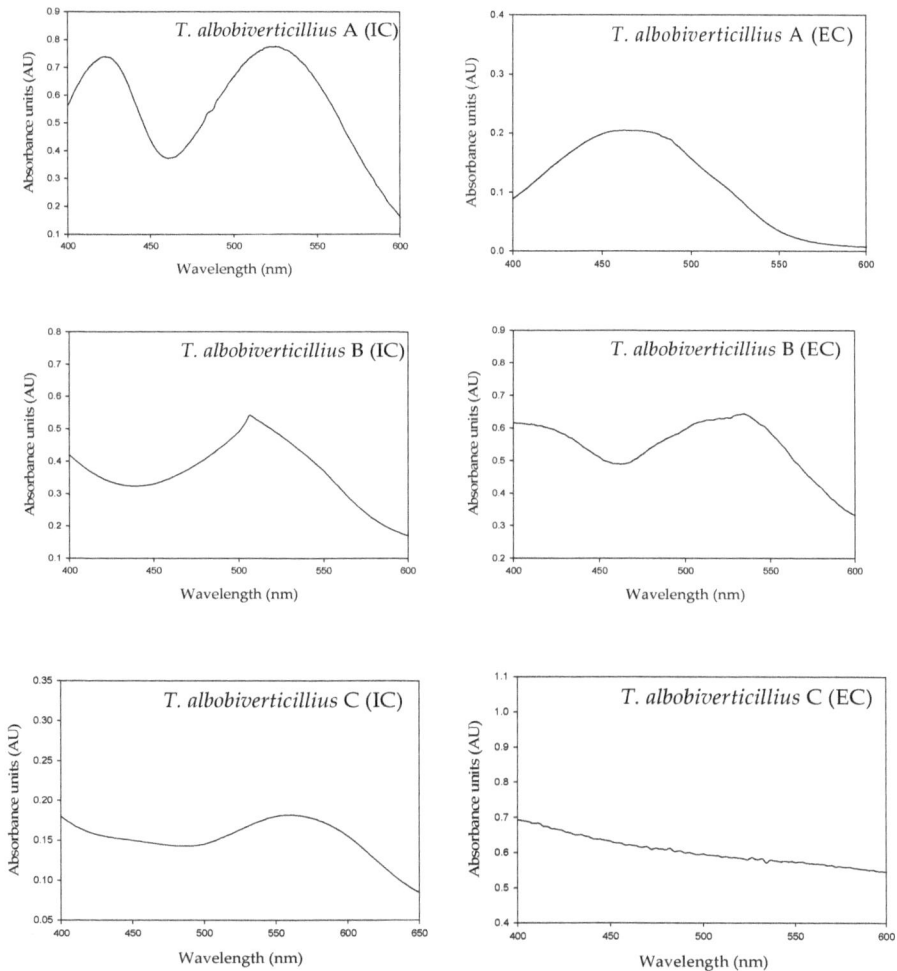

(c)

Figure 4. Intracellular (IC) and extracellular (EC) UV-visible spectra of: (**a**) *Aspergillus creber* A and B; (**b**) *Penicillium viticola* B and C; and (**c**) *Talaromyces albobiverticillius* A–C cultures in potato dextrose broth (7 days).

However, slight variations were noticed between extra- and intracellular liquids: in *A. creber* A as an example, extracellular maximum absorbance was around 470 nm (yellow-orange hue) instead of 550 nm (red shade) for intracellular liquid (Figure 4a). These slight variations however indicate that intra- and extracellular solutions may contain different assortments of colored compounds, in different proportions, resulting in different hues (Figure 5).

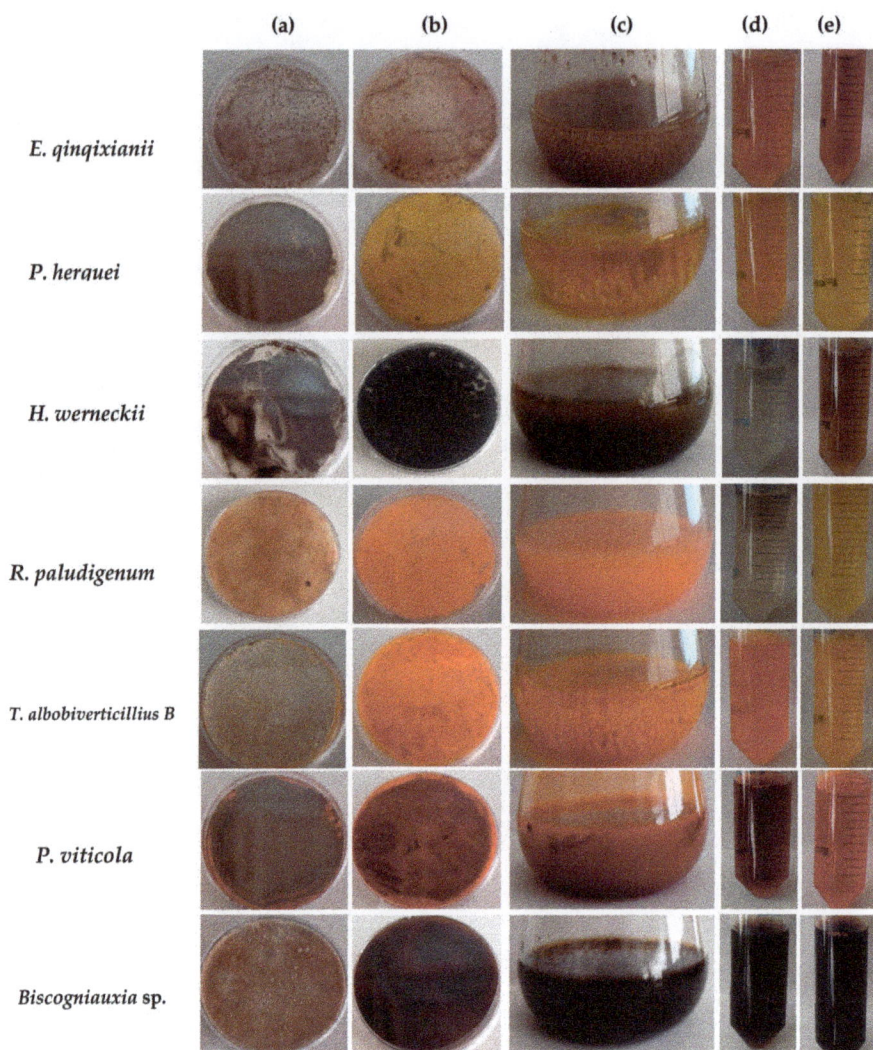

Figure 5. Colors observed in different fungal species: (**a**) obverse face on PDA; (**b**) reverse face on PDA; (**c**) culture in PDB (seven days); (**d**) extract of intracellular pigments (Ethanol/water 50/50) (IC); and (**e**) filtrate from liquid culture (EC).

Differences were also observed among the spectral profiles of different isolates belonging to the same species. As shown from the intracellular profiles of *A. creber* A and B, *P. viticola* B and C, and *T. albobiverticillius* A–C (Figure 4a–c, respectively, and Table 4), maximal absorbance areas differed in the visible region (510–560 nm for *P. viticola* B and 420–450 nm for *P. viticola* C; and 422–525 nm for *T. albobiverticillius* A, 500 nm for *T. albobiverticillius* B and 520–580 nm for *T. albobiverticillius* C), but, for *A. creber* A and B, the spectra looked similar (Figure 4a,b). Similar variation was stated between the extracellular profiles.

These results clearly imply that isolates from a same species produce and secrete different pigments and therefore have different behavior towards colored compound production.

Table 4. Summary of main peaks (λ_{max}) noticed in 10-days old culture of *Talaromyces albobiverticillius* isolates A–C cultivated in liquid medium (potato dextrose broth).

T. albobiverticillius	Sample	Peaks in the UV Region (nm)		Peaks in the Visible Region (nm)	
		200–250	250–300	300–400	>400
A	IC	235	286	362	422, 425, 511, 525
	EC		265	365	458, 469.8, 480
B	IC	232	268, 292		410, 440, 460, 500
	EC		288		412, 524, 532
C	IC	222	283	385	520–580
	EC		283	370, 385	436

3.4. Evaluation of Intracellular and Extracellular Contents in Pigments

The amount of pigments in IC and EC solutions, expressed in mg eq. Purpurin L^{-1}, are presented in Figure 6.

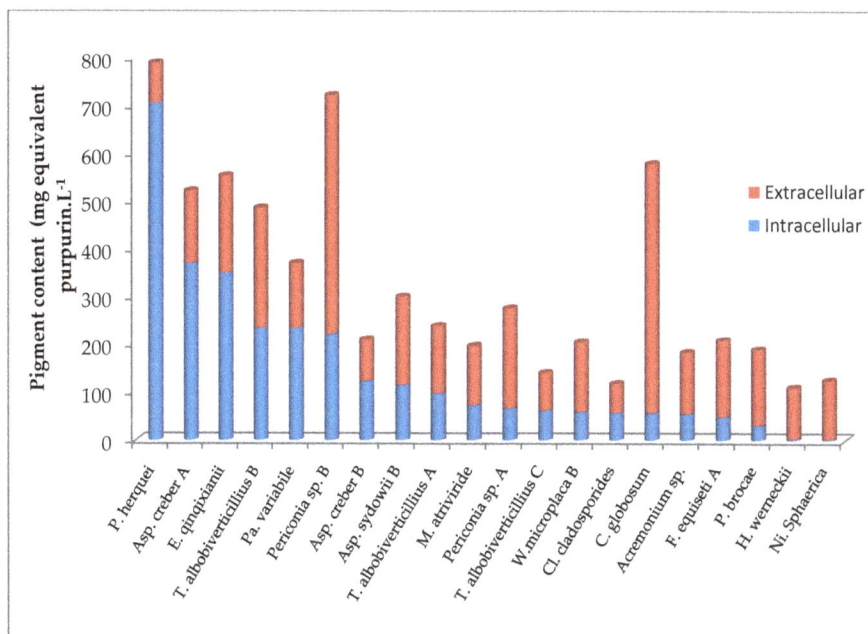

Figure 6. Colored compounds in biomass (intracellular, IC) and culture filtrate (extracellular, EC) for 20 isolates of marine derived fungi isolated around La Réunion Island, in mg eq. purpurin L^{-1} of culture medium (potato dextrose broth, absorbance at 254 nm).

In regard with the diversity of isolates and colored compounds involved in this study, instead of the intensity of the color, the values express the global amount of polyketides compounds produced by each isolate, hues ranging from pale green, light yellow to dark red or maroon.

In the extracellular samples, the maximum amount was produced by *Chaetomium globosum* (521.44 mg equivalent purpurin L^{-1}), followed by *Periconia* sp. B (498.39 mg equivalent purpurin L^{-1}). For intracellular samples, the maximum levels were measured for *P. herquei*, *A. creber* A, *E. qinqixianii* with 704.55, 371.18 and 350.93 mg equivalent purpurin L^{-1}, respectively.

The amount of intracellular content was significantly lower than the one of the extracellular content in this population (n = (20,20), V = 10, P = 6 × 10^{-4}). However, looking at each isolate separately, the amount of intracellular pigments was significantly higher than the extracellular one for *P. herquei* (704.55 vs. 84 mg equivalent purpurin L^{-1}), and *A. creber* A (371.18 vs. 151.11) and B (125.25 vs. 86.35), and *E. qinqixianii* (350.93 vs. 202.74).

4. Discussion

4.1. Biodiversity of Marine-Derived Fungi around La Réunion Island

From the sampling in La Réunion Island marine biotopes, 31 different species distributed in 25 genera were identified as pigment producers. The identification of isolates collected in coral reefs and lava flows of La Réunion Island coincides with identifications conducted from various marine environments. Indeed, the majority of the studied fungi, such as those sampled from north of the Indian Ocean, belong to the phylum Ascomycetes. The fungi of the genus *Aspergillus*, particularly *A. sydowii*, are also found in the Caribbean corals (*Porites lobata*), Polynesia, and in sediments off the coast of India [46,47]. *Penicillium citrinum* was isolated from the red algae *Actinotrichia fragilis*, from sponges, and the species was also found on other substrates such as hard substrate or water [48–51]. The genera *Penicillium*, *Cladosporium*, *Chaetomium*, and *Fusarium*, and species *Nigrospora oryzae* and *Hortea werneckii*, have been identified in marine sediments collected at different depths in the central basin of the Indian Ocean and considered to be coral pathogens [8,52]. Fungi, namely *Alternaria* sp., *Acremonium* sp. and *Rhodotorula mucilaginosa*, were isolated from salt lakes in Antarctica, as were *P. chrysogenum* and *P. crustosum* [53]. *Rhodotorula mucilaginosa* was also found in the sediments of central Indian basin [52].

The diversity within the isolated fungal population was crucial while comparing the ability of pigment production [54]. However, in our samples, the highest diversity of pigmented fungi was revealed from the water column (13 species) and from hard substrates (limestone or lava flow) (11 species). If the water column can be suspected of carrying a multitude of fungal propagules originating from terrestrial environments, hard substrates are probably more representative of marine and marine-derived biodiversity.

Our study demonstrates that the living coral *Pocillopora* sp. shelters fungi from the genera *Aspergillus* (*A. creber*, *A. sydowii* and *Eurotium amstellodami*) and *Penicillium* (*P. viticola*), as well as *Hypocrea koningii* and *Acremonium* sp. Widely disseminated on land, this mainly saprophytic genus *Acremonium* sp., has already been isolated from marine environments (sea fans, sea water, sea cucumbers, and intertidal sediment samples) [55–57]. These fungi, were extracted from the inner parts of the coral structure. They are then supposed to be at least endophytic species for this coral genus.

The coral rubbles (dead part of corals) from our samples contained colored *Penicillium* (*P. herquei*) or related species *T. albobiverticillius* as well as an isolate of *Chaetomium globosum*. *Chaetomium globosum* is a common fungal species from soil and environment.

Most of the fungi we identified can also be found on land, in soil, on plants or insects, but some of them have rarely been isolated from marine environments such as *Whalleya microplaca*, *Biscogniauxia* sp., *Paraconiothyrium variabile*, *Myrothecium atroviride*, *Nectria haematococca*, *Peyronellaea glomerata*, *Epicoccum sorghi*, *Sporisorium exsertum* and *Periconia* sp.. From our study, the genera *Aspergillus* and *Penicillium* or the close ones such as *Talaromyces*, *Emericella* or *Eurotium* (from the Trichocomaceae family) are much more diverse than others in these tropical marine biotopes (12 different species), and are represented in several types of samples and locations. These aerobic and xerophilic species are well-known for populating dry and/or salty biotopes. However, their ability to subsist or develop underwater, with widely varying oxygenation conditions is less known. These cosmopolitan fungi are well-known to produce a wide range of secondary metabolites such as polyketide-based pigments in solid and liquid cultures. Overall, in our study, some fungal species (*T. albobiverticillius* or *N. haematococca*) were identified from different types of samples in the same area. Some others (*A. sydowii*, *P. viticola*) appeared in separate locations. These fungi found in several sample types and/or in different locations may be considered as frequent in marine environment around La Réunion Island.

4.2. Qualitative Aspect of the Pigment Production

For marine-derived isolates, two statuses lead to particular behaviors and products: the challenge of facing unusual living conditions (exogenous fungi) and the use of specific procedures naturally adapted to the marine niches (for instance fungal endophytes of marine microorganisms, i.e., indigenous micromycetes, naturally selected by aquatic environments).

Overall, in unusual biotopes (sometimes extreme), the fungal species with pigmented cell walls (in the spores and/or mycelium), are clearly able to tolerate dehydration-hydration cycles or high solar radiations, better than the moniliaceous fungi, whose cells are devoid of pigments. These aromatic compounds, as melanin, sporopollenin (brown product of oxydative polymerization of β-carotene) or cycloleucomelone (terphenylquinone), often show significant antioxidant activities, and are bound to protect the biological structures, giving them an excellent durability and a high potential for survival in hostile environments [58,59].

From the available literature, the microorganisms of the genus *Trichoderma* are frequent in marine environments and some terrestrial strains are able to produce anthraquinone-like compounds [60]. Indeed, isolates of the family Hypocreaceae (*Trichoderma*, *Hypocrea* and *Acremonium*) are also represented in our study and exhibit orange to purple hues. Some strains of the common soil fungus *Cladosporium cladosporioides*, also isolated from our samples with green shades, have already been studied for their production of intracellular melanin [61].

The most important colored compounds produced by *Aspergillus* and *Emericella* species are respectively, hydroxyanthraquinones and azaphilone pigments, exhibiting a very wide range of hues. Furthermore, *A. sydowii* and *Eurotium amstelodami* isolated from La Réunion Island showed red and yellow colors respectively, as produced by their terrestrial counterparts [62].

Penicillium species and related ones seem to adjust easily to multiple conditions and to be a source of original compounds as they appear among the most chemically inventive fungi. In *Penicillium* and *Talaromyces* species, polyketide-based pigments are also very common, and, particularly, the azaphilones, such as the derivatives of monascorubrin and rubropunctatin [63]. *Monascus*-like azaphilone pigments such as N-glutarylmonascorubramine, N-glutarylrubropunctamine, monascorubramine homologues PP-V [(10Z)-12-carboxyl-monascorubramine] and PP-R [(10Z)-7-(2-hydroxyethyl)-monascorubramine] are frequently identified in their cultures [64,65]. However, the commercial production of red anthraquinoid pigments (Arpink Red™, Natural Red™) has already been carried out with *P. oxalicum var. armeniaca* [1]. The most common hues produced by both genera include yellow, red, orange and reddish-brown. Nevertheless, it was found that the yellow pigments seem predominant in most of the *Penicillium* species, while *Talaromyces* species mainly produce red pigments with few synthesizing yellow compounds of azaphilone series [66]. The colored molecules sometimes demonstrate mycotoxic activities such as rubratoxins A and B, rugulovasins and luteoskyrin [67].

Some strains of the widespread *Acremonium* sp. produce the yellow oosporein (chaetomidin) (biquinone, benzoquinone) and also some toxic compounds as diterpene glycosides [68].

Chaetomium globosum, isolated from the coral rubbles biosynthesizes maroon pigments in the culture conditions of our experiment. Many members of the family produce metabolites with antifungal properties. *C. globosum* is already known to biosynthesize yellow azaphilones named chaetoviridins (A–D), antifungal compounds involved in the induction of chlamydospores-like cells [69]. It also produces nitrogenous azaphilones (4'-epi-N-2-hydroxyethyl-azachaetoviridin A, and N-2-butyric-azochaetoviridin E) and isochromophilone XIII, with orange to red hues. Some strains generate pigmented chaetoglobins, chaetoglobosins, chaetomugilins, and seco-chaetomugilins, while others can secrete a purple pigment called cochliodinol [70–73].

Associated with lava flows, *Fusarium equiseti* belongs to a group of widespread plant pathogens, but marine-derived *Fusarium* strains are also frequent in mangroves or associated with marine organisms. These are already known to produce original colored anthraquinoid compounds (5-acetyl-2-methoxy-1,4,6-trihydroxy-anthraquinone;6,8-dimethoxy-1-methyl-2-(3-oxobutyl)-anthraquinone and fusaquinones) [19]. Among the *Fusarium* secondary metabolites, numerous polyketide pigments have already been identified, such as naphthoquinone pigments which are the most abundant (bikaverin, nor-bikaverin, javanicin, anhydrojavanicin, fusarubin, anhydrofusarubin, bostrycoidin, and novarubin) and the hydroxyanthraquinones emodin, physcion, dermolutein, chrysophanol, erythroglaucin, dermocybin, dermorubin, tritisporin, cynodontin, helminthosporin or aurofusarin (review in [19,21]). All these molecules develop a palette of colors, ranging from yellow to purple or brown. Some species are also able to produce orange carotenoids (neurosporaxanthin by *F. fujikuroi*) [74]. The putative carcinogen, fusarin C, apicidin F, fujikurins, the perithecal pigments fusarubins as well as the mycelial pigment bikaverin are also produced in the family.

From our work, *Periconia* sp. A isolate produced an impressive violet hue in PDB culture. *Periconia* is a cosmopolitan genus, often found in soil, and decaying herbs and forages. Some *Periconia* strains were nevertheless identified from marine environments (*P. abyssa* (deep sea), *P. byssoides* (sea slug *Aplysia kurodai*)) [75–77]. They attract interest because of the production of promising anti-cancer drugs, such as the carbosugar pericosine A. Some strains may produce an unidentified hepatoxin.

4.3. Quantitative Aspect of Pigment Production

As a promising factor, several of the marine-derived fungi isolated in this study had the ability to grow and biosynthesize pigments in unsalted synthetic conditions (e.g., Czapek Dox medium, PDB). During the period of fermentation, the pigment production started between Day 1 and Day 4 for the majority of isolates such as *Aspergillus*, *Eurotium*, *Fusarium*, *Nigrospora*, *Pencillium* and *Talaromyces*. For some fungi, the detection of the pigment production was notably delayed (e.g., *Acremonium*, *Epicoccum*, or *Myrothecium*). This might be due to the low level of pigment producing ability of the fungi or due to unfavorable environmental conditions for pigment production such as pH, temperature, nutrient sources, osmolarity and illumination conditions [78].

Considering the visual observation of pigment color in flasks and the respective UV-visible spectra, fungi belonging to the same species may produce different colored mixtures (e.g., *Aspergillus creber* A and B or *Talaromyces alboverticillius* A–C). They may then belong to different varieties and thus produce pigments of distinct natures. The slight variations observed between intra- and extra-cellular solutions also indicate that the solutions may contain different assortments of colored compounds, in different proportions, resulting in different hues.

From these findings, it is understood that a higher quantity of pigments has been mainly purified from extracellular filtrates in a significant manner (11/21 isolates). In our experimental conditions, the maximum pigment production was obtained in the extracellular samples for *C. globosum* and *Periconia* sp. B. On average, the values measured in the cells were significantly lower; indicating that pigments secretion in the liquid medium seems a widespread behavior in the conditions of the experiment. Only the isolate *P. herquei* had a very high level of intracellular pigment biosynthesis (704.55 mg equivalent purpurin L^{-1}). Nevertheless, for high intracellular pigment production from biomass, *A. creber* A and *E. qinqixianii* present a true production potential. On the other hand, the extraction of intracellular colored compounds appeared sometimes not completely effective. The fungal biomass was still colored even after extraction. The efficiency of the extraction process could probably be improved to recover higher pigment quantities from intracellular samples [79].

This work highlights different behaviors of fungal isolates towards the secretion of colored molecules compared to internal storage. Anyway, the production of secondary metabolites often occurs after fungal growth has ceased, as a result of nutrient limitation coupled with excess carbon availability. This makes it possible to manipulate their formation [80,81].

5. Conclusions

Marine and marine-derived fungi are promising resources for the production of new metabolites of interest, and, among them, pigments are attractive [82–84]. The potential of marine-derived microorganisms to produce unique and original molecules may come from specific metabolic or genetic adaptation appearing to meet very specific combinations of physical and chemical parameters (high salinity, low O_2 penetration, low temperature, limited light access and high pressure). Based on this statement, our study explores, for the first time, the biodiversity of fungi from marine environments around La Réunion Island, Indian Ocean, along with the ability of the isolates to produce pigments. The potentiality of these marine derived isolates to secrete pigments or to concentrate colored compounds inside the cells was highlighted. Several isolates collected from lava flows, hard substrates sediments and corals (living or dead) turned out to be the interesting producers of intense colors on PDA culture medium. The main types identified, *Aspergillus*, *Penicillium* and related genera, are also found in other marine regions (such as Polynesia or along the coast of India). However, a great biodiversity (31 species) emphasizes the range of possible hues and molecules susceptible to be isolated. The majority of the isolates, probably marine optional, may also be able to grow in synthetic media, devoid of sea salts and may show the competence of producing pigments in an industrial scale. The most promising pigmented products, probably of intense red or purple hues, which seem to consist in mixtures, will be subjected to purification and further analyses by analytical techniques such as liquid chromatography–mass spectrometry/time-of-flight (LC-MS/TOF) and Nuclear Magnetic Resonance (NMR). The interesting isolates will also be subjected to further analyses to determine their ability as antibiotics or for enzyme production.

Supplementary Materials: The following are available online at www.mdpi.com/2309-608X/3/3/36/s1.

Acknowledgments: The authors are grateful to Regional Council of La Réunion Island for financial support. Thanks are also given to BIOLAVE program and Quod Jean Pascal.

Author Contributions: Mireille Fouillaud conceived, designed and performed the experiments. Mireille Fouilaud and Pascale Cuet collected the samples. Melissa Llorente, Hélène Magalon and Mekala Venkatachalam performed the molecular analysis and analyzed the genetic data. Mireille Fouillaud, Mekala Venkatachalam and Laurent Dufossé contributed to write the paper. The authors would also like to thank Cathie Milhau from ESIROI and Patricia Clerc from LCSNSA, of Université de La Réunion, for their logistic and technical help; and Gary Mares for his timely help on data analysis.

Conflicts of Interest: The authors declare no conflict of interest.

References

1. Dufossé, L.; Galaup, P.; Yaron, A.; Arad, S.M.; Blanc, P.; Chidambara Murthy, K.N.; Ravishankar, G.A. Microorganisms and microalgae as sources of pigments for food use: A scientific oddity or an industrial reality? *Trends Food Sci. Technol.* **2005**, *16*, 389–406. [CrossRef]

2. Mayer, A.M.; Rodriguez, A.D.; Taglialatela-Scafati, O.; Fusetani, N. Marine pharmacology in 2009–2011: Marine compounds with antibacterial, antidiabetic, antifungal, anti-inflammatory, antiprotozoal, antituberculosis, and antiviral activities; affecting the immune and nervous systems, and other miscellaneous mechanisms of action. *Mar. Drugs* **2013**, *11*, 2510–2573. [PubMed]

3. Mapari, S.A.S.; Nielsen, K.F.; Larsen, T.O.; Frisvad, J.C.; Meyer, A.S.; Thrane, U. Exploring fungal biodiversity for the production of water-soluble pigments as potential natural food colorants. *Curr. Opin. Biotechnol.* **2005**, *16*, 231–238. [CrossRef] [PubMed]

4. Hohmann, C.; Schneider, K.; Bruntner, C.; Irran, E.; Nicholson, G.; Bull, A.T.; Jones, A.L.; Brown, R.; Stach, J.E.; Goodfellow, M.; et al. Caboxamycin, a new antibiotic of the benzoxazole family produced by the deep-sea strain *Streptomyces* sp. Ntk 937. *J. Antibiot.* **2009**, *62*, 99–104. [CrossRef] [PubMed]

5. Costantino, V.; Fattorusso, E.; Mangoni, A.; Perinu, C.; Cirino, G.; De Gruttola, L.; Roviezzo, F. Tedanol: A potent anti-inflammatory ent-pimarane diterpene from the caribbean sponge *Tedania ignis*. *Bioorg. Med. Chem.* **2009**, *17*, 7542–7547. [CrossRef] [PubMed]

6. Bonugli-Santos, R.C.; dos Santos Vasconcelos, M.R.; Passarini, M.R.Z.; Vieira, G.A.L.; Lopes, V.C.P.; Mainardi, P.H.; dos Santos, J.A.; de Azevedo Duarte, L.; Otero, I.V.R.; da Silva Yoshida, A.M.; et al. Marine-derived fungi: Diversity of enzymes and biotechnological applications. *Front. Microbiol.* **2015**, *6*, 269. [CrossRef] [PubMed]
7. Panno, L.; Bruno, M.; Voyron, S.; Anastasi, A.; Gnavi, G.; Miserere, L.; Varese, G.C. Diversity, ecological role and potential biotechnological applications of marine fungi associated to the seagrass *Posidonia oceanica*. *New Biotechnol.* **2013**, *30*, 685–694. [CrossRef] [PubMed]
8. Cathrine, S.J.; Raghukumar, C. Anaerobic denitrification in fungi from the coastal marine sediments off Goa, India. *Mycol. Res.* **2009**, *113*, 100–109. [CrossRef] [PubMed]
9. Holguin, G.; Vazquez, P.; Bashan, Y. The role of sediment microorganisms in the productivity, conservation, and rehabilitation of mangrove ecosystems: An overview. *Biol. Fertil. Soils* **2001**, *33*, 265–278. [CrossRef]
10. Bugni, T.S.; Ireland, C.M. Marine-derived fungi: A chemically and biologically diverse group of microorganisms. *Nat. Prod. Rep.* **2004**, *21*, 143–163. [CrossRef] [PubMed]
11. Jones, E.B.G.; Pang, K.L. *Marine Fungi: And Fungal-Like Organisms*; De Gruyter: Berlin, Germany, 2012.
12. Kathiresan, K.; Bingham, B.L. Biology of mangroves and mangrove ecosystems. In *Advances in Marine Biology*; Southward, A., Young, C., Fuiman, L., Tyler, P., Eds.; Academic Press: San Diego, CA, USA, 2001; Volume 40, pp. 81–251.
13. Kohlmeyer, J.; Kohlmeyer, E. *Marine Mycology*; Elsevier Inc.: London, UK, 1979; p. 704.
14. Jones, E.B.G. Marine fungi: Some factors affecting biodiversity. *Fungal Divers.* **2000**, *4*, 53–73. [CrossRef]
15. Saleem, M.; Nazir, M. Bioactive natural products from marine-derived fungi: An update. In *Studies in Natural Products Chemistry*; Atta-ur-Rahman, Ed.; Elsevier: Amsterdam, The Netherlands, 2015; Volume 45, pp. 297–361.
16. Imhoff, J.F. Natural products from marine fungi—Still an underrepresented resource. *Mar. Drugs* **2016**, *14*, 19. [CrossRef] [PubMed]
17. Debashish, G.; Malay, S.; Barindra, S.; Joydeep, M. Marine enzymes. *Adv. Biochem. Eng./Biotechnol.* **2005**, *96*, 189–218.
18. Zhang, C.; Kim, S.K. Application of marine microbial enzymes in the food and pharmaceutical industries. *Adv. Food Nutr. Res.* **2012**, *65*, 423–435. [PubMed]
19. Fouillaud, M.; Venkatachalam, M.; Girard-Valenciennes, E.; Caro, Y.; Dufossé, L. Anthraquinones and derivatives from marine-derived fungi: Structural diversity and selected biological activities. *Mar. Drugs* **2016**, *14*, 64. [CrossRef] [PubMed]
20. Fouillaud, M.; Venkatachalam, M.; Girard-Valenciennes, E.; Caro, Y.; Dufossé, L. Marine-derived fungi producing red anthraquinones: New resources for natural colors? In Proceedings of the 8th International Conference of Pigments in Food, "Coloured Foods for Health Benefits", Cluj-Napoca, Romania, 28 June—1 July 2016.
21. Caro, Y.; Venkatachalam, M.; Lebeau, J.; Fouillaud, M.; Dufossé, L. Pigments and colorants from filamentous fungi. In *Fungal Metabolites*; Merillon, J.-M., Ramawat, G.K., Eds.; Springer: Cham, Switzerland, 2016; pp. 1–70.
22. Ebel, R. Natural product diversity from marine fungi. In *Comprehensive Natural Products II: Chemistry and Biology*; Mander, L., Liu, H.-W., Eds.; Elsevier: Oxford, UK, 2010; Volume 2, pp. 223–262.
23. Calvo, A.M.; Wilson, R.A.; Bok, J.W.; Keller, N.P. Relationship between secondary metabolism and fungal development. *Microbiol. Mol. Biol. Rev.* **2002**, *66*, 447–459. [CrossRef] [PubMed]
24. Margalith, P. *Pigment Microbiology*; Springer: London, UK; New York, NY, USA, 1992; p. 156.
25. Demain, A.L.; Fang, A. The natural functions of secondary metabolites. In *History of Modern Biotechnology I*; Fiechter, A., Ed.; Springer: Berlin/Heidelberg, Germany, 2000; pp. 1–39.
26. Réunion's Coral Reef. Available online: https://en.wikipedia.org/wiki/R%C3%A9union%27s_coral_reef (accessed on 1 May 2017).
27. Peyrot-Clausade, M.; Chazottes, V.; Pari, N.; Peyrot-Clausade, M.; Chazottes, V.; Pari, N. Bioerosion in the carbonate budget of two indo-pacific reefs: La Réunion (Indian Ocean) and moorea (Pacific Ocean). *Bull. Geol. Soc. Denmark* **1999**, *1999*, 1–30.

28. Conand, C.; Chabanet, P.; Cuet, P.; Letourneur, Y. The carbonate budget of a fringing reef in La Reunion Island (Indian Ocean): Sea urchin and fish bioerosion and net calcification. In Proceedings of the 8th International Coral Reef Symposium, Panama City, Panama, 24–29 June 1997; Lessios, H.A., Macintyre, I.G., Eds.; pp. 953–958.

29. Naim, O.; Cuet, P.; Mangar, V. Coral reefs of the mascarene archipelago. In *Coral Reefs of the Indian Ocean: Their Ecology and Conservation*; McClanahan, T.R., Sheppard, C., Obura, D.O., Eds.; Oxford University Press: New York, NY, USA, 2000; pp. 353–381.

30. Turner, J.; Klaus, R. Coral reefs of the mascarenes, western indian ocean. *Philos. Trans. R. Soc. Lond. A Math. Phys. Eng. Sci.* **2005**, *363*, 229–250. [CrossRef] [PubMed]

31. Montaggioni, L.; Faure, G. *Les Récifs Coralliens des Mascareignes (Océan Indien)*; Université Française de l'Océan Indien, Centre Universitaire de La Réunion: Réunion, France, 1980; p. 151.

32. Sanders, E.R. Aseptic laboratory techniques: Plating methods. *J. Vis. Exp. JoVE* **2011**, *63*, e3064. [CrossRef] [PubMed]

33. Jong, S.; Dugan, F.; Edwards, M. *ATCC Filamentous Fungi*, 19th ed.; Rockville, MD American Type Culture Collection: Manassas, VA, USA, 1996.

34. Dahmen, H.; Staub, T.; Schwinn, F. Technique for long-term preservation of phytopathogenic fungi in liquid nitrogen. *Phytopathology* **1983**, *73*, 241–246. [CrossRef]

35. Knebelsberger, T.; Stoger, I. DNA extraction, preservation, and amplification. *Methods Mol. Biol.* **2012**, *858*, 311–338. [PubMed]

36. Toju, H.; Tanabe, A.S.; Yamamoto, S.; Sato, H. High-coverage its primers for the DNA-based identification of ascomycetes and basidiomycetes in environmental samples. *PLoS ONE* **2012**, *7*, e40863. [CrossRef] [PubMed]

37. Samson, R.A.; Visagie, C.M.; Houbraken, J.; Hong, S.B.; Hubka, V.; Klaassen, C.H.W.; Perrone, G.; Seifert, K.A.; Susca, A.; Tanney, J.B.; et al. Phylogeny, identification and nomenclature of the genus aspergillus. *Stud. Mycol.* **2014**, *78*, 141–173. [CrossRef] [PubMed]

38. Samson, R.A.; Yilmaz, N.; Houbraken, J.; Spierenburg, H.; Seifert, K.A.; Peterson, S.W.; Varga, J.; Frisvad, J.C. Phylogeny and nomenclature of the genus talaromyces and taxa accommodated in penicillium subgenus biverticillium. *Stud. Mycol.* **2011**, *70*, 159–183. [CrossRef] [PubMed]

39. White, T.J.; Bruns, T.; Lee, S.; Taylor, J. Amplification and direct sequencing of fungal ribosomal RNA genes for phylogenetics. *PCR Protoc. Guide Methods Appl.* **1990**, *18*, 315–322.

40. Romanelli, A.M.; Sutton, D.A.; Thompson, E.H.; Rinaldi, M.G.; Wickes, B.L. Sequence-based identification of filamentous basidiomycetous fungi from clinical specimens: A cautionary note. *J. Clin. Microbiol.* **2010**, *48*, 741–752. [CrossRef] [PubMed]

41. Sutton, D.A.; Marín, Y.; Thompson, E.H.; Wickes, B.L.; Fu, J.; García, D.; Swinford, A.; de Maar, T.; Guarro, J. Isolation and characterization of a new fungal genus and species, aphanoascella galapagosensis, from carapace keratitis of a galapagos tortoise (chelonoidis nigra microphyes). *Med. Mycol.* **2013**, *51*, 113–120. [CrossRef] [PubMed]

42. Zhou, H.; Li, Y.; Tang, Y. Cyclization of aromatic polyketides from bacteria and fungi. *Nat. Prod. Rep.* **2010**, *27*, 839–868. [CrossRef] [PubMed]

43. Caro, Y.; Anamale, L.; Fouillaud, M.; Laurent, P.; Petit, T.; Dufosse, L. Natural hydroxyanthraquinoid pigments as potent food grade colorants: An overview. *Nat. Prod. Bioprospect.* **2012**, *2*, 174–193. [CrossRef]

44. Machatová, Z.; Barbieriková, Z.; Poliak, P.; Jančovičová, V.; Lukeš, V.; Brezová, V. Study of natural anthraquinone colorants by epr and uv/vis spectroscopy. *Dyes Pigments* **2016**, *132*, 79–93. [CrossRef]

45. Geyer, C.J. Fuzzy *p*-Values and Ties in Nonparametric Tests. Avaliable online: http://www.stat.umn.edu/geyer/fuzz (accessed on 30 June 2017).

46. Golubic, S.; Radtke, G.; Le Campion-Alsumard, T. Endolithic fungi in marine ecosystems. *Trends Microbiol.* **2005**, *13*, 229–235. [CrossRef] [PubMed]

47. Priess, K.; Le Campion-Alsumard, T.; Golubic, S.; Gadel, F.; Thomassin, B. Fungi in corals: Black bands and density-banding of *Porites lutea* and *P. lobata* skeleton. *Mar. Biol.* **2000**, *136*, 19–27. [CrossRef]

48. Nicoletti, R.; Trincone, A. Bioactive compounds produced by strains of penicillium and *Talaromyces* of marine origin. *Mar. Drugs* **2016**, *14*, 37. [CrossRef] [PubMed]

49. Tsuda, M.; Kasai, Y.; Komatsu, K.; Sone, T.; Tanaka, M.; Mikami, Y.; Kobayashi, J.i. Citrinadin a, a novel pentacyclic alkaloid from marine-derived fungus *Penicillium citrinum*. *Org. Lett.* **2004**, *6*, 3087–3089. [CrossRef] [PubMed]

50. Malmstrøm, J.; Christophersen, C.; Frisvad, J.C. Secondary metabolites characteristic of *Penicillium citrinum*, *Penicillium steckii* and related species. *Phytochemistry* **2000**, *54*, 301–309. [CrossRef]
51. Endo, A.; Kuroda, M.; Tsujita, Y. Ml-236a, ml-236b, and ml-236c, new inhibitors of cholesterogensis produced by *Penicillium citrinum*. *J. Antibiot.* **1976**, *29*, 1346–1348. [CrossRef] [PubMed]
52. Singh, P.; Raghukumar, C.; Verma, P.; Shouche, Y. Assessment of fungal diversity in deep-sea sediments by multiple primer approach. *World J. Microbiol. Biotechnol.* **2012**, *28*, 659–667. [CrossRef] [PubMed]
53. Brunati, M.; Rojas, J.L.; Sponga, F.; Ciciliato, I.; Losi, D.; Gottlich, E.; de Hoog, S.; Genilloud, O.; Marinelli, F. Diversity and pharmaceutical screening of fungi from benthic mats of antarctic lakes. *Mar. Genom.* **2009**, *2*, 43–50. [CrossRef] [PubMed]
54. Yahr, R.; Schoch, C.L.; Dentinger, B.T. Scaling up discovery of hidden diversity in fungi: Impacts of barcoding approaches. *Philos. Trans. R. Soc. B* **2016**, *371*, 20150336. [CrossRef] [PubMed]
55. An, X.; Feng, B.-M.; Chen, G.; Chen, S.-F.; Wang, H.-F.; Pei, Y.-H. Isolation and identification of two new compounds from marine-derived fungus *Acremonium fusidioides* rz01. *Chin. J. Nat. Med.* **2016**, *14*, 934–938. [CrossRef]
56. Afiyatullov, S.S.; Kalinovsky, A.I.; Antonov, A.S.; Zhuravleva, O.I.; Khudyakova, Y.V.; Aminin, D.L.; Yurchenko, A.N.; Pivkin, M.V. Isolation and structures of virescenosides from the marine-derived fungus *Acremonium striatisporum*. *Phytochem. Lett.* **2016**, *15*, 66–71. [CrossRef]
57. Gallardo, G.L.; Butler, M.; Gallo, M.L.; Rodríguez, M.A.; Eberlin, M.N.; Cabrera, G.M. Antimicrobial metabolites produced by an intertidal acremonium furcatum. *Phytochemistry* **2006**, *67*, 2403–2410. [CrossRef] [PubMed]
58. Hiort, J.; Maksimenka, K.; Reichert, M.; Perovic-Ottstadt, S.; Lin, W.H.; Wray, V.; Steube, K.; Schaumann, K.; Weber, H.; Proksch, P.; et al. New natural products from the sponge-derived fungus *Aspergillus niger*. *J. Nat. Prod.* **2004**, *67*, 1532–1543. [CrossRef] [PubMed]
59. Pagano, M.C.; Rosa, L.H. Fungal molecular taxonomy. In *Fungal Biomolecules*; John Wiley & Sons, Ltd.: Chichester, UK, 2015; pp. 311–321.
60. Slater, G.; Haskins, R.; Hogge, L.; Nesbitt, L. Metabolic products from a *Trichoderma viride* pers. Ex fries. *Can. J. Chem.* **1967**, *45*, 92–96. [CrossRef]
61. Duran, N.; Teixeira, M.F.; De Conti, R.; Esposito, E. Ecological-friendly pigments from fungi. *Crit. Rev. Food Sci. Nutr.* **2002**, *42*, 53–66. [CrossRef] [PubMed]
62. Butinar, L.; Frisvad, J.C.; Gunde-Cimerman, N. Hypersaline waters–a potential source of foodborne toxigenic *Aspergilli* and *Penicillia*. *FEMS Microbiol. Ecol.* **2011**, *77*, 186–199. [CrossRef] [PubMed]
63. Woo, P.C.; Lam, C.W.; Tam, E.W.; Lee, K.C.; Yung, K.K.; Leung, C.K.; Sze, K.H.; Lau, S.K.; Yuen, K.Y. The biosynthetic pathway for a thousand-year-old natural food colorant and citrinin in *Penicillium marneffei*. *Sci. Rep.* **2014**, *4*, 6728. [CrossRef] [PubMed]
64. Arai, T.; Koganei, K.; Umemura, S.; Kojima, R.; Kato, J.; Kasumi, T.; Ogihara, J. Importance of the ammonia assimilation by *Penicillium purpurogenum* in amino derivative monascus pigment, PP-V, production. *AMB Express* **2013**, *3*, 19. [CrossRef] [PubMed]
65. Ogihara, J.; Kato, J.; Oishi, K.; Fujimoto, Y. Pp-r, 7-(2-hydroxyethyl)-monascorubramine, a red pigment produced in the mycelia of *Penicillium* sp. AZ. *J. Biosci. Bioeng.* **2001**, *91*, 44–47. [CrossRef]
66. Frisvad, J.C.; Yilmaz, N.; Thrane, U.; Rasmussen, K.B.; Houbraken, J.; Samson, R.A. *Talaromyces atroroseus*, a new species efficiently producing industrially relevant red pigments. *PLoS ONE* **2013**, *8*, e84102. [CrossRef] [PubMed]
67. Yilmaz, N.; Houbraken, J.; Hoekstra, E.S.; Frisvad, J.C.; Visagie, C.M.; Samson, R.A. Delimitation and characterisation of *Talaromyces purpurogenus* and related species. *Persoonia* **2012**, *29*, 39–54. [CrossRef] [PubMed]
68. Thomson, R.H. *Naturally Occurring Quinones IV: Recent Advances*; Blackie Academic & Professional: London, UK; New York, NY, USA, 1997.
69. Takahashi, M.; Koyama, K.; Natori, S. Four new azaphilones from *Chaetomium globosum* var. Flavo-viridae. *Chem. Pharm. Bull.* **1990**, *38*, 625–628. [CrossRef]
70. McMullin, D.R. Structural Characterization of Secondary Metabolites Produced by Fungi Obtained from Damp Canadian Building. Ph.D. Thesis, Ottawa-Carleton University, Ottawa, ON, Canada, 2008.

71. McMullin, D.R.; Sumarah, M.W.; Miller, J.D. Chaetoglobosins and azaphilones produced by Canadian strains of *Chaetomium globosum* isolated from the indoor environment. *Mycotoxin Res.* **2013**, *29*, 47–54. [CrossRef] [PubMed]

72. Ming Ge, H.; Yun Zhang, W.; Ding, G.; Saparpakorn, P.; Chun Song, Y.; Hannongbua, S.; Xiang Tan, R. Chaetoglobins A and B, two unusual alkaloids from endophytic *Chaetomium globosum* culture. *Chem. Commun.* **2008**, 5978–5980. [CrossRef] [PubMed]

73. Brewer, D.; Jerram, W.A.; Taylor, A. The production of cochliodinol and a related metabolite by chaetomium species. *Can. J. Microbiol.* **1968**, *14*, 861–866. [CrossRef] [PubMed]

74. Prado-Cabrero, A.; Schaub, P.; Diaz-Sanchez, V.; Estrada, A.F.; Al-Babili, S.; Avalos, J. Deviation of the neurosporaxanthin pathway towards beta-carotene biosynthesis in *Fusarium fujikuroi* by a point mutation in the phytoene desaturase gene. *FEBS J.* **2009**, *276*, 4582–4597. [CrossRef] [PubMed]

75. Kohlmeyer, J. New genera and species of higher fungi from the deep sea (1615–5315 m). *Revue de Mycologie* **1977**, *41*, 189–206.

76. Usami, Y.; Ichikawa, H.; Arimoto, M. Synthetic efforts for stereo structure determination of cytotoxic marine natural product pericosines as metabolites of *Periconia* sp. From sea hare. *Int. J. Mol. Sci.* **2008**, *9*, 401–421. [CrossRef] [PubMed]

77. Dighton, J.; White, J.F. *The Fungal Community: Its Organization and Role in the Ecosystem*, 3rd ed.; CRC Press: Boca Raton, FL, USA, 2005.

78. Ogbonna, C.N. Production of food colourants by filamentous fungi. *Afr. J. Microbiol. Res.* **2016**, *10*, 960–971.

79. Kaufmann, B.; Christen, P. Recent extraction techniques for natural products: Microwave-assisted extraction and pressurised solvent extraction. *Phytochem. Anal. PCA* **2002**, *13*, 105–113. [CrossRef] [PubMed]

80. Debbab, A.; Aly, A.H.; Proksch, P. Bioactive secondary metabolites from endophytes and associated marine derived fungi. *Fungal Divers.* **2011**, *49*, 1–12. [CrossRef]

81. Gunatilaka, A.L.; Wijeratne, E.K. *Natural Products from Bacteria and Fungi*; Elsevier: Amsterdam, The Netherlands, 2000.

82. Kim, S.-K. *Marine Microbiology: Bioactive Compounds and Biotechnological Applications*; Wiley & Sons: Weinheim, Germany, 2013; p. 550.

83. Kjer, J.; Debbab, A.; Aly, A.H.; Proksch, P. Methods for isolation of marine-derived endophytic fungi and their bioactive secondary products. *Nat. Protoc.* **2010**, *5*, 479–490. [CrossRef] [PubMed]

84. Rateb, M.E.; Ebel, R. Secondary metabolites of fungi from marine habitats. *Nat. Prod. Rep.* **2011**, *28*, 290–344. [CrossRef] [PubMed]

Journal of
Fungi

MDPI

Article

Combinatorial Biosynthesis of Novel Multi-Hydroxy Carotenoids in the Red Yeast *Xanthophyllomyces dendrorhous*

Hendrik Pollmann [1], Jürgen Breitenbach [1], Hendrik Wolff [2], Helge B. Bode [2,3] and Gerhard Sandmann [1,*]

[1] Biosynthesis Group, Molecular Biosciences, Fachbereich Biowissenschaften, Goethe Universität Frankfurt, Frankfurt am Main 60438, Germany; pollmann@bio.uni-frankfurt.de (H.P.); Breitenbach@em.uni-frankfurt.de (J.B.)
[2] Merck Stiftungsprofessur für Molekulare Biotechnologie, Fachbereich Biowissenschaften, Goethe Universität Frankfurt, Frankfurt am Main 60438, Germany; wolff@bio.uni-frankfurt.de (H.W.); h.bode@bio.uni-frankfurt.de (H.B.B.)
[3] Buchmann Institute for Molecular Life Sciences (BMLS), Goethe Universität Frankfurt, Max-von-Laue-Strasse 15, Frankfurt am Main 60438, Germany
* Correspondence: sandmann@bio.uni-frankfurt.de; Tel.: +49-69-798-29625; Fax: +49-69-798-29600

Academic Editors: Laurent Dufossé, Yanis Caro and Mireille Fouillaud
Received: 13 January 2017; Accepted: 14 February 2017; Published: 22 February 2017

Abstract: The red yeast *Xanthophyllomyces dendrorhous* is an established platform for the synthesis of carotenoids. It was used for the generation of novel multi oxygenated carotenoid structures. This was achieved by a combinatorial approach starting with the selection of a β-carotene accumulating mutant, stepwise pathway engineering by integration of three microbial genes into the genome and finally the chemical reduction of the resulting 4,4'-diketo-nostoxanthin (2,3,2',3'-tetrahydroxy-4,4'-diketo-β-carotene) and 4-keto-nostoxanthin (2,3,2',3'-tetrahydroxy-4-monoketo-β-carotene). Both keto carotenoids and the resulting 4,4'-dihydroxy-nostoxanthin (2,3,4,2',3',4'-hexahydroxy-β-carotene) and 4-hydroxy-nostoxanthin (2,3,4,2'3'-pentahydroxy-β-carotene) were separated by high-performance liquid chromatography (HPLC) and analyzed by mass spectrometry. Their molecular masses and fragmentation patterns allowed the unequivocal identification of all four carotenoids.

Keywords: carotenoid biosynthesis; 4,4'-dihydroxy-nostoxanthin; 4,4'-diketo-nostoxanthin; genetic engineering; HPLC separation; MS-MS spectra; *Xanthophyllomyces dendrorhous*

1. Introduction

Xanthophyllomyces dendrorhous (with *Phaffia rhodozyma* as its anamorphic state) is a basidiomycetous red yeast that accumulates astaxanthin [1], which is a unique feature among fungi [2]. This carotenoid is formed via the mevalonate pathway starting with a condensation of two molecules of geranylgeranyl pyrophosphate, a 4-step desaturation, cyclization and a final 4-ketolation plus 3-hydroxylation [3,4]. In contrast to other organisms, only three genes are involved in the whole pathway from phytoene to astaxanthin, which facilitates genetic modification of carotenoid biosynthesis in *X. dendrorhous*. This yeast has the potential to be engineered as a cell-factory for the production of industrially valuable carotenoids [5]. Tools and techniques for genetic manipulations of *X. dendrorhous* are available [6], including several integrative transformation plasmids based on four different selection markers [7,8]. It is also advantageous for the development of high-yield carotenoid producers since a carotenoid pathway that can be manipulated is already established, a carotenoid storage system exists and a very active acetyl-CoA metabolism can be utilized. The published genomic sequence of *X. dendrorhous* CBS6938 [9] is also very helpful for genetic engineering of the carotenoid pathway.

The potential of *X. dendrorhous* for the production of economically interesting carotenoids like astaxanthin for feed and zeaxanthin as a nutraceutical for eye care has been demonstrated. The highest production of astaxanthin was reached by combining classical mutagenesis with genetic pathway engineering [7]. Starting from a high-yield astaxanthin mutant, genes of three limiting enzymes were over-expressed, enhancing metabolite flow toward carotenoid biosynthesis and into the astaxanthin pathway [7].

Zeaxanthin is another carotenoid of interest as a nutraceutical, which is important for protection of our vision. For the engineering of a zeaxanthin producing strain, a mutant with inactive astaxanthin synthase accumulating β-carotene [10,11] was used to extend the carotenoid pathway to zeaxanthin by expression of a bacterial β-carotene hydroxylase gene and engineering of an enhanced metabolite flow into the carotenoid pathway [12].

Hydroxylated carotenoids such as the dihydroxy compound zeaxanthin [13], tetrahydroxy-β-carotene derivative nostoxanthin [14] and various other hydroxylated acyclic carotenoids [15] have been generated in *Escherichia coli* by combination of carotenogenic genes from different organisms. It has been shown that not only zeaxanthin but also several other hydroxyl derivatives have superior antioxidative activity. In our engineering approach with *X. dendrorhous*, we utilized β-carotene mutants to integrate hydroxylase and ketolase genes from bacteria and algae and finally chemically reduced the resulting tetrahydroxy-monoketo and tetrahydroxy-diketo products to a pentahydroxy and a hexahydroxy β-carotene derivative, respectively.

2. Materials and Methods

2.1. Strains and Cultivation

The β-carotene accumulation mutant of *X. dendrorhous* strain PR1-104 was generated by ethyl-methane sulphonate mutagenesis and selection for carotenoid content [10]. This mutant and transformants were grown as shaking cultures (50 mL in 500 mL baffled Erlenmeyer flasks, 180 rpm) over 7 days in YM medium (0.3% yeast extract, 0.3% malt extract, 0.5% peptone, 1.0% glucose) with white light illumination. For selection of transformants on agar plates, geneticin (G418 sulfate, 100 μg/mL), hygromycin (60 μg/mL) or nourseothricin (30 μg/mL) were added. *Escherichia coli* strains DH5α and JM110 used for genetic manipulations and generation of carotenoid standards were grown in LB medium containing ampicillin (100 μg/mL) and chloramphenicol (34 μg/mL) according to the plasmids involved.

2.2. Plasmid Construction and Transformantion of X. dendrorhous, Combinatorial Biosynthesis of Carotenoid Standards in E. coli

The integrative plasmids pPR2TNo-crtZo [12] and pPRcDNA1bkt830 [11] were previously described. Details on the origin of genes and the primers for the amplification of the *X. dendrorhous* transformation plasmids are shown in Table 1. For the construction of plasmid pPR2TNH-crtG, the *crtG* gene from plasmid pUC-Bre-O11 [16] was amplified, which also generated an *Eco*RI and *Xho*I restriction site (Table 1), and was cloned into the *Xcm*I site of the *E. coli* plasmid pMon38201 by its a-overhang [17]. From there, the *crtG* gene was cut out with *Eco*RI and *Xho*I and ligated into pUC8ΔEcoRI-HNNH [7] for fusion with the promoter and terminator of the glyceraldehyde phosphate dehydrogenase gene from *X. dendrorhous*. Then, the whole cassette was cut out by restriction with *Hind*III and ligated into the *Hind*III site of pPR2TNH [7], yielding pPR2TNH-crtG. Transformation of *X. dendrorhous* was performed by electroporation as described in Visser et al. [6] with 10 μg plasmid DNA. After phenol-chloroform purification, plasmid DNA was linearized by digestion with *Sfi*I.

Carotenoid standards were generated in *E. coli* by combinatorial transformation of plasmids pACCAR25ΔcrtX [18] plus pUC-Bre-O11 [16] for nostoxanthin, caloxanthin and zeaxanthin, pACCAR25ΔcrtX [18] plus pCRBKT [19] for canthaxanthin, echinenone and β-carotene and pACCAR25ΔcrtX plus pPEU30crtO [20] for the synthesis of 4-keto-zeaxanthin.

Table 1. Genes and oligonucleotides for PCR amplification of the *Xanthophyllomyces dendrorhous* transformation plasmids.

Plasmid	Primers
pPRcDNA1bkt830	β-Carotene 4-ketolase gene *bkt* from *Haematococcus pluvialis* (D45881) Primers: 5′-ATATGAATTCATGCACGTCGCATCGGCACT-3′ (forward) 5′-TATACTCGAGTCATGCCAAGGCAGGCACCAG-3′ (reverse)
pPR2TNH-crtG	β-Carotene 2-hydroxylase gene *crtG* from *Brevundimonas* SD212 [16] (AB181388) Primers: 5′-GGAATTCATGTTGAGGGATCTGCTCATC-3′ (forward) 5′-GCTCGAGTCACCGAAGAGGCGCTGAG-3′ (reverse)
pPR2TNo-crtZo	β-Carotene 3-hydroxylase gene *crtZ* from *Brevundimonas* SD212 [21]), codon optimized (KX063854) Sequences: 5′-GAATTCATGGCTTGGCTCACCTGGA-3′ (start) 5′-ATCTTCTTCTTCTGGAGCT TGAGTCGAC-3′ (end)

2.3. Carotenoid Extraction, Purification and Chemical Derivatization

For carotenoid extraction, 20 mg of freeze-dried *X. dendrorhous* cells were mixed with 500 μL glass beads (0.25–0.5 mm), 675 μL methanol and 75 μL of a 60% KOH solution for saponification and broken in a cell mill (Retsch MM 400) for 8 min at a frequency of 30 per second and heated for 20 min at 60 °C. Samples containing keto carotenoid were extracted without addition of KOH. After partitioning of the carotenoid extract into 50% diethyl ether in petrol (bp 40–60 °C), the upper phase was collected and dried in a stream of nitrogen. Carotenoids were quantified from three independently grown cultures.

For the purification of multi-hydroxy keto carotenoids, the extract of PR1-104-ZGbkt was fractionated by TLC on activated silica plates developed with toluene/ethyl acetate/methanol (65:35:5, by volume). The bands with an R_f value of 0.20 and 0.25 were collected and extracted with acetone.

The reduction of carotenoid ketones dissolved in ethanol to the corresponding alcohol was performed according to Eugster [22] with $NaBH_4$. The reaction mixture was transferred to 65% diethyl ether in petrol (bp 40–60 °C) and the upper phase collected and dried in stream of nitrogen. The same procedure was applied to canthaxanthin, yielding 4,4′-diketo-β-carotene together with 4-HO-4′-keto-β-carotene and to echinenone yielding 4-HO-β-carotene.

2.4. HPLC and HR-ESI-MS Analysis

HPLC analysis was performed on three different HPLC systems. For separation of the non-ketolated multi-hydroxy carotenoids, system I with a 15 × 0.4 cm Nucleosil 100 C18, 3 μm column and acetonitrile (ACN)/methanol/2-propanol (85:10:5, by volume) plus 3% H_2O [23] was used as mobile phase with a flow rate of 0.8 mL/min at 10 °C. System II was employed for the separation of the hydroxy-keto carotenoids and their reduction products on a 25 cm C30 RP, 3 μm column (YMC, Wilmington, NC, USA) with a mobile phase of 3% methyl tertiary-butyl ether in methanol for 48 min followed by an increase to 20% with a flow rate of 0.8 mL/min at 20 °C. This system was also used for the quantification of the keto derivatives. HR-ESI-MS analysis was carried out in system III on a 2.1 mm × 50 mm ACQUITY UPLC BEH C18, 1.7 μm column. A binary gradient was applied with ACN (+0.1% formic acid) and H_2O (+0.1% formic acid) for 12 min in the following steps: 0–2 min 5% ACN , 2–2.5 min 40% ACN, 2.5–4 min 40% ACN, 4–14 min 40%–95% at a flow rate of 0.4 mL/min at 40 °C. This HPLC system was coupled to an Impact II QTOF (Bruker) mass spectrometer using Na-formiate as an internal calibration standard [24]. Carotenoids were detected in a positive ion mode with scanning range from 100–1200 *m/z*. Optical spectra were recorded online with a photodiode array detector 994 (Waters, Milford, CT, USA). Carotenoid standards for identification were generated in *E. coli* by the combination of different *crt* genes as previously described [25].

3. Results

Although fungi are in general unable to synthesize zeaxanthin or other hydroxy-carotenoids, formation of the 3,3′-dihydroxy-β-carotene can be engineered into *X. dendrorhous* [12]. This potential was extended for the synthesis of other derivatives with up to six hydroxyl groups. As outlined in Figure 1, this was achieved by a strategy involving the use of a β-carotene accumulating mutant of *X. dendrorhous*, its consecutive transformation with three microbial transgenes and finally the reduction of the 4- and 4′-keto groups. Figure 2 shows the HPLC analysis to identify some of the β-carotene-derived oxo compounds from the transgenic lines. Transformation with a bacterial 3-hydroxylase gene changed the PR1-104 from a β-carotene accumulator (trace A) to a transformant synthesizing zeaxanthin together with small amounts of the intermediate β-cryptoxanthin (trace B). In a second transformation step, with a bacterial 2-hydroxylase gene, the line PR1-104-ZG was obtained in which β-carotene was converted to nostoxanthin (trace C) as indicated by the nostoxanthin standard in trace D. Oxygenation of nostoxanthin was increased by a third transformation step with an algal 4-ketolase gene. In the resulting transformant PR1-104-ZGbkt, several carotenoids were detected. For base-line HPLC separation of these very polar oxo carotenoids, system I had to be changed to system II (Figure 2, trace E). In addition to the expected nostoxanthin-derived keto carotenoids, other carotenoids such as β-carotene (peak 7) and its keto derivatives echinenone (peak 6) and canthaxanthin (peak 5) as well as 4-keto-zeaxanthin (peak 4) and nostoxanthin (peak 3) were generated and could be identified with standard carotenoids (Figure 2, traces F and G). Compounds **1** and **2** of trace E are highly polar diketo and mono keto derivatives, respectively, as indicated by their optical absorbance spectra exhibiting a typical shape for keto carotenoids and an absorbance maximum of 475 or 468 nm (Figure 3).

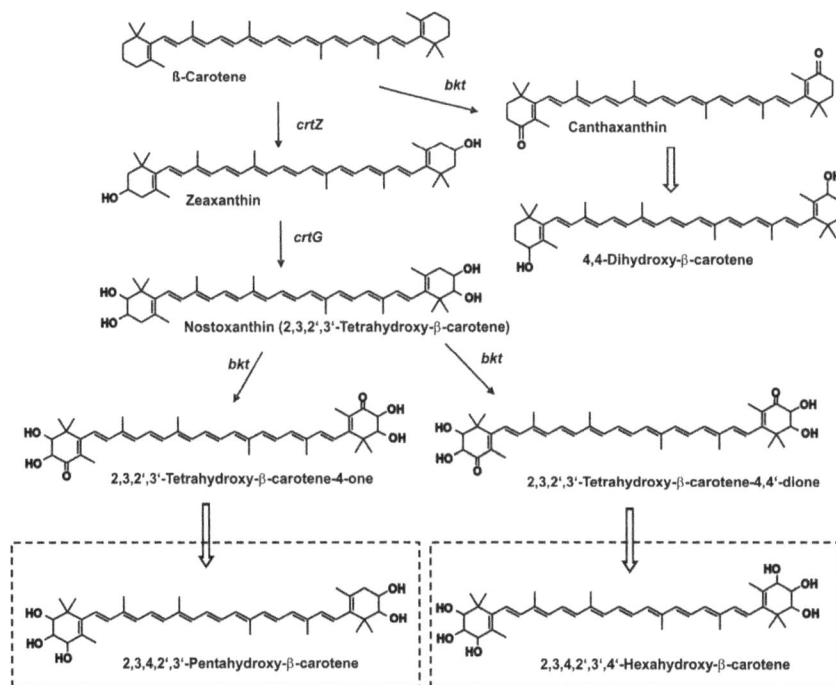

Figure 1. Pathway construction by genetic engineering of *Xanthophyllomyces dendrorhous* for the synthesis of multi-oxygenated β-carotene derivatives. Open arrows indicate chemical reduction. Novel carotenoid structures are boxed.

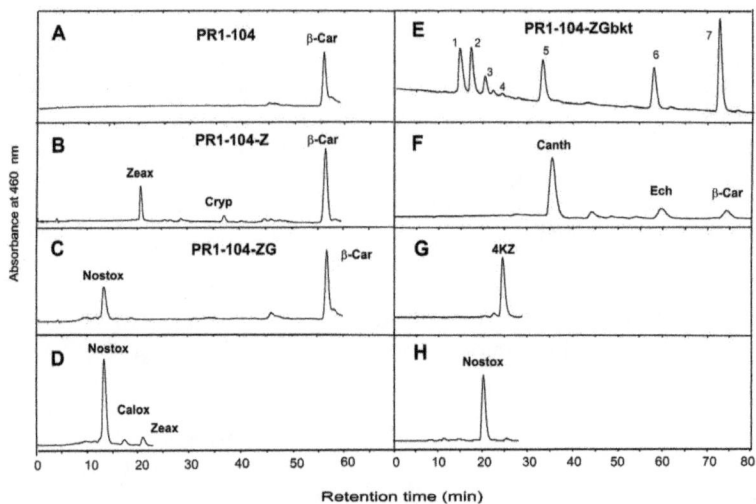

Figure 2. HPLC separation of hydroxy and keto carotenoids from *Xanthophyllomyces dendrorhous* lines obtained by consecutive transformation with different *trans* genes. Traces **A–D** separated in HPLC system I, **E–H** in system II. Standards are shown in traces **D, F–H**. Abbreviations (and assignment of corresponding peaks): β-Car β-carotene (7), Zeax zeaxanthin, Cryp β-cryptoxanthin, Nostox nostoxanthin (3), Calox caloxanthin, Canth canthaxanthin (5), Ech echinenone (6), 4KZ 4-keto-zeaxanthin (4).

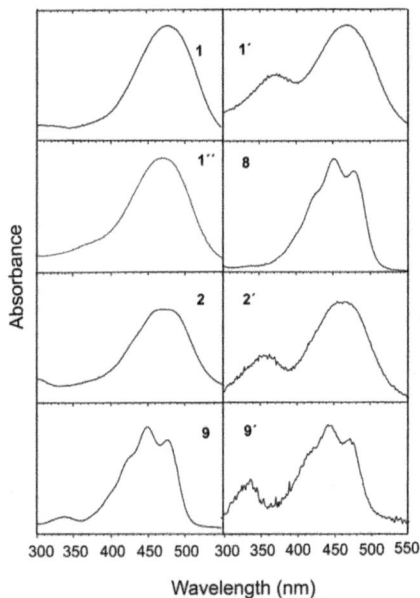

Figure 3. Absorbance spectra of hydroxy-keto carotenoids before (compounds **1** and **2** with *cis* isomers **1′, 1″** and **2′**) and after reduction (compounds **8** and **9** with *cis* isomer **9′**).

The carotenoid extract from transformant PR1-104-ZGbkt was reduced and separated by HPLC (Figure 4A). New peaks 8 to 12 emerged, which were not present in the non-reduced sample. Peaks 10 to 12 could be identified by chromatography of standards as 4,4′-dihydroxy-β-carotene (trace C), 4-hydroxy-4′-keto-β-carotene (trace B) and 4-hydroxy-β-carotene (trace D). Peaks 8 and 9 exhibited similar spectra with absorbance maxima and shoulder values at 424, 450 and 477 nm (Figure 3). Both compounds were obtained when isolated individual compounds from the non-reduced fraction were reduced. Isolated compound **1** changed into compound **8** and **2** into **9** (Figure 4, right part). The absorbance spectra of the all-*trans* isomers **1, 2, 8** and **9** together with individual *cis* carotenoids indicated with same primed number are shown in Figure 3.

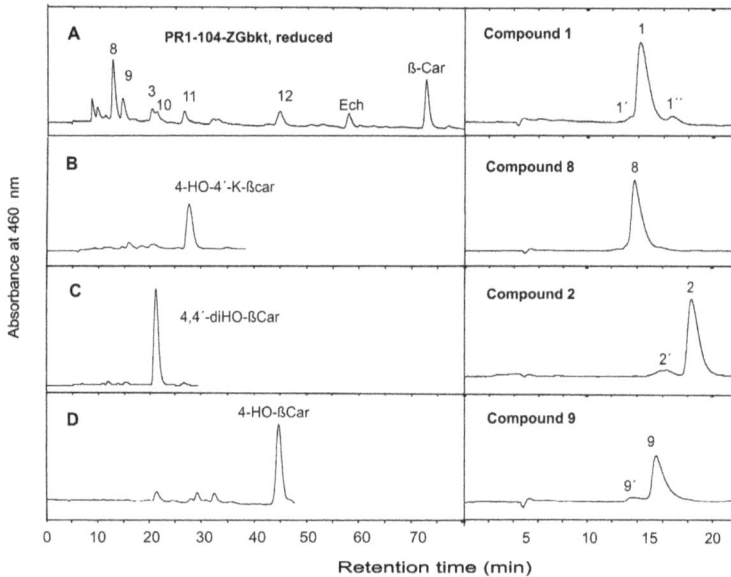

Figure 4. HPLC separation of hydroxy carotenoids after reduction (**A–D**) and isolated peaks 1 and 2 from Figure 2 and their reduction products in system II. Standards are shown in traces **B–D**. Abbreviations (and assignment of corresponding peaks): β-Car, β-carotene; Ech, echinenone; 4-HO-4′-K-βcar, 4-hydroxy-4′-keto-β-carotene (11); 4,4′-diHO-βCar, 4,4′-dihydroxy-β-carotene (10); 4-HO-β-Car, 4-hydroxy-β-carotene (12).

Final identification of compounds **1, 2, 8** and **9** was performed using high resolution mass spectrometry (Figure 5). Compound **1** showed a molecular mass of 628.3745 Da (Table 2) and fragments of M-17, M-17-17 and M-92 (Figure 5A). This molecular mass identifies compound **1** as 4,4′-diketo-nostoxanthin. Its reduction product, compound **8**, with a molecular mass of 632.4077 Da (Table 2) is regarded as 4,4′-dihydroxy-nostoxanthin. Instead of the M-17 and M-17-17 fragments of compound **1**, the fragments M-18 and M-18-18 were present in its mass spectrum. (Figure 5B). Therefore, compound **8** is regarded as 4,4′-dihydroxy-nostoxanthin. Compound **2** exhibits a fragmentation pattern in which the M-17-17 fragment of compound **1** is replaced by M-17-18 (Figure 5C). It has a molecular mass of 614.3947 Da (Table 2). This identifies compound **2** as 4-keto-nostoxanthin. Upon its reduction, the molecular mass increased to 616.4102 Da (Table 2). The M-92 fragment is retained but instead of M-17, an M-18 fragment and instead of M-17-18, an M-18-18 fragment appears (Figure 5D). These features of compound **9** correspond to 4-hydroxy-nostoxanthin. Common to the keto carotenoid **1** and **2** is a prominent peak at 147.12, which is much less pronounced in the reduced compounds **8** and **9**. The concentrations of these keto carotenoids in *X. dendrorhous* were determined as 37.4 ± 1.1 (μg/g dw) for 4,4′-diketo-nostoxanthin and 48.8 ± 1.4 (μg/g dw) for 4-keto-nostoxanthin. Their chemical reduction was almost complete (Figure 4A), which implies similar concentrations for 4,4′-dihydroxy-nostoxanthin and 4-hydroxy-nostoxanthin, respectively.

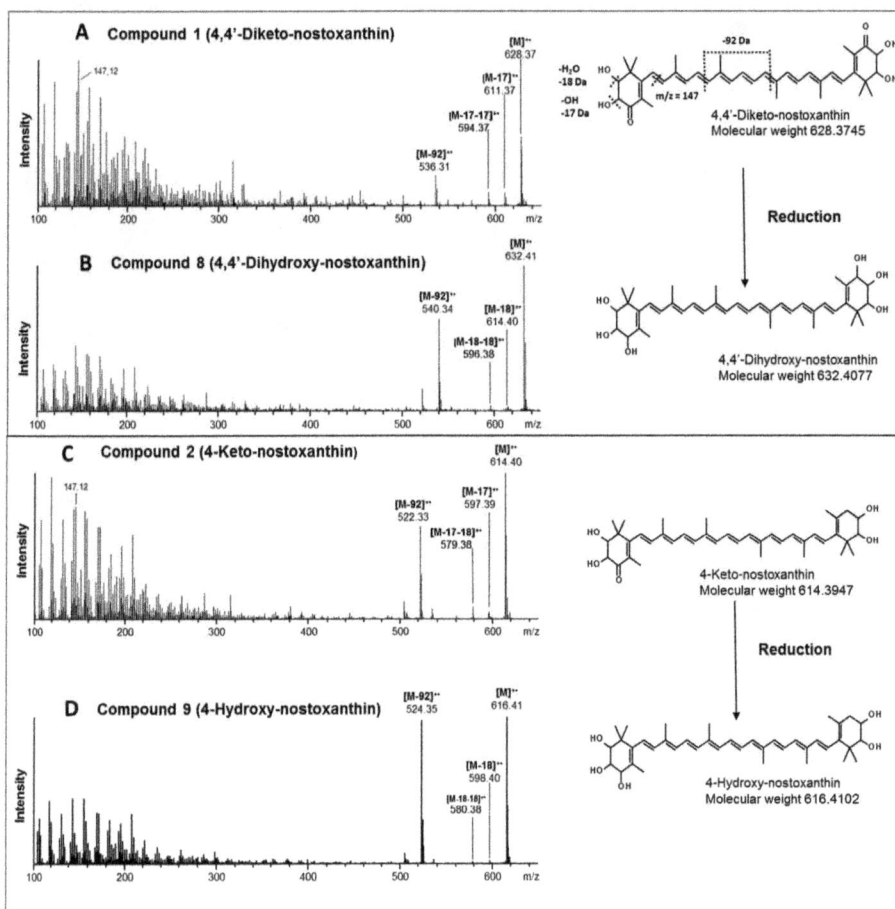

Figure 5. (A–D) MS-MS analysis of de-novo generated hydroxyl and keto-hydroxy β-carotene derivatives before and after reduction. The structures are indicated including the fragmentation pattern exemplified for 4-keto-nostoxanthin.

Table 2. Overview of identified multi-oxygenated nostoxanthins using UPLC-ESI-HRMS.

Compound	Trivial Name	Sum Formula [M]+	Calculated *m/z*	Detected *m/z*	Mass Error [Δppm]
1	4,4′-Diketo-nostoxanthin	$C_{40}H_{52}O_6$	628.3758	628.3745	1.3
2	4-Keto-nostoxanthin	$C_{40}H_{54}O_5$	614.3966	614.3947	1.9
8	4,4′-Dihydroxy-nostoxanthin	$C_{40}H_{56}O_6$	632.4071	632.4077	0.6
9	4-Hydroxy-nostoxanthin	$C_{40}H_{56}O_5$	616.4122	616.4102	2.0

4. Discussion

Production of carotenoids by genetically engineered yeasts proved to be a promising alternative to chemical synthesis or extraction from plants [26]. The red yeast *X. dendrorhous* is the most versatile host with the highest carotenoid yield among fungi [5]. It is possible to construct and implement pathways to different carotenoid structures into this yeast. As an example, the synthesis to multi-oxygenated carotenoids was chosen in this publication (Figure 1) to demonstrate the potential of *X. dendrorhous* as a production platform for complex carotenoid structures. This was possible by extension of the pathway from accumulating β-carotene. The step-by-step transformation resulted in intermediary lines accumulating zeaxanthin or nostoxanthin as major carotenoids (Figure 2). Nostoxanthin is a carotenoid found in cyanobacteria [27] and was accumulated in recombinant *Escherichia coli* [14]. A final *X. dendrorhous* line PR1-104-ZGbkt transformed with three microbial genes produced 4-keto-nostoxanthin and 4,4′-diketo-nostoxanthin (Figure 2). Both carotenoids are extremely rare in nature and have been identified before only from two bacteria *Brevundimonas* SD212 and *Rhizobium lupine* [28,29]. By reduction of both keto carotenoids isolated from our line PR1-104-ZGbkt, the novel carotenoids 4-hydroxy-nostoxanthin and 4,4′-dihydroxy-nostoxanthin were obtained (Figure 4). Some of the isolated oxo carotenoids separated into several geometrical isomers on the C30 column (Figure 3). In each case, the all-*trans* isomer dominated. The isomers **1′** and **2′** showed a *cis* peak at 370 nm and **9′** at 330 nm. According to their height in relation to the dominating absorbance maximum, their position in front of the all-*trans* isomer on a C30 column and in comparison to astaxanthin for the keto derivatives [6] and to zeaxanthin for 4-hydroxy-nostoxanthin [30], these isomers are most likely 13-*cis*. In contrast, **1″** without a *cis* peak may be a 9-*cis* 4,4′-diketo-nostoxanthin isomer.

The keto carotenoids and their reduction products were identified by high resolution mass spectrometry (Figure 5). For all of them, the correct molecular masses and the typical prominent fragments were obtained (Table 2). As indicated in the right part of Figure 5A, fragment M-92 originates from an in chain elimination of a toluene unit and is an indication of the central polyene chain [31]. All analyzed carotenoids show the elimination of hydroxyl groups either as water (M-18) or as neutral loss of 17 Da. In addition, an intensified peak was found at 147.12 Da in the spectra of the mono and diketo derivaties (Figure 5A,C). This is typical for a dehydrated 4-keto-β-ionone ring with cleavage of the C7,8 bond [31].

Our combined approach of mutant selection, genetic engineering and chemical modification is set as a general example of how novel carotenoids can be generated in *X. dendrorhous*. For the production of 4-keto-nostoxanthin and 4,4′-diketo-nostoxanthin, it is a proof of concept, which also indicates how to improve their yields in future studies. Judging from the relative low conversion of β-carotene to zeaxanthin (Figure 2B) and complete conversion of zeaxanthin to nostoxanthin (Figure 2C), the 3-hydroxylation step is regarded limiting in the formation of ketolated nostoxanthin. Formation of ketolated β-carotene derivatives echinenone and canthaxanthin demonstrate that the ketolation step is not limited (Figure 2E). It has previously been shown that conversion rates of transgenic reactions in *X. dendrorhous* are dependent on the number of *trans* gene copies integrated into the genome [32]. Therefore, either transformation with a plasmid carrying two copies of the 3-hydroxylase gene *crtZ* as demonstrated by Pollmann et al. [12] or repeated transformation with the *crtZ* gene is a promising way to improve intermediate conversion to the end product. In addition, overall carotenoid synthesis can be enhanced up to 3-fold by improvement of precursor supply in combination with increased flux into the carotenoid pathway [12] and in combination with a high carotenoids producing *X. dendrorhous* mutant, a total increase of carotenoid formation of up to 90-fold can be achieved [7].

Acknowledgments: This work was supported as the ERA-IB project PROCAR through the German Federal Ministry of Education and Research (BMBF) (FKZ 031A569) and in part by COST Action CA15136. The high resolution ESI-MS system was partially funded by the DFG.

Author Contributions: Hendrik Pollmann and Gerhard Sandmann designed the experiments and carried out the HPLC analysis, Jürgen Breitenbach constructed the transformation plasmids, Hendrik Pollmann carried out transformation and cultivation of *X. dendrorhous* including sample preparation, Hendrik Wolff and Helge B. Bode performed high resolution MS-MS measurement and data analysis. All authors commented on the completed manuscript.

Conflicts of Interest: The authors declare no conflict of interest.

References

1. Andrewes, A.G.; Starr, M.P. (3R,3′R)-Astaxanthin from the yeast *Phaffia rhodozyma*. *Phytochemistry* **1976**, *15*, 1009–1011. [CrossRef]
2. Sandmann, G.; Misawa, N. Fungal carotenoids. In *The Mycota X; industrial application*; Osiewacz, H.D., Ed.; Springer: Berlin, Germany, 2002; pp. 247–262.
3. Rodriguez-Saiz, M.; de la Fuente, J.L.; Barredo, J.L. *Xanthophyllomyces dendrorhous* for the industrial production of astaxanthin. *Appl. Microbiol. Biotechnol.* **2010**, *88*, 645–658. [CrossRef] [PubMed]
4. Schmidt, I.; Schewe, H.; Gassel, S.; Jin, C.; Buckingham, J.; Hümbelin, M.; Sandmann, G.; Schrader, J. Biotechnological production of astaxanthin with *Phaffia rhodozyma/Xanthophyllomyces dendrorhous*. *Appl. Microbiol. Biotechnol.* **2010**, *8*, 555–571. [CrossRef] [PubMed]
5. Sandmann, G. Carotenoids of biotechnological importance. In *Biotechnology of Isoprenoids, Advances in Biochemical Engineering/Biotechnology*; Schrader, J., Bohlmann, J., Eds.; Springer: Berlin/Heidelberg, Germany, 2014; Volume 148, pp. 449–467.
6. Visser, H.; Sandmann, G.; Verdoes, J.C. Xanthophylls in fungi: metabolic engineering of the astaxanthin biosynthetic pathway in *Xantophyllomyces. dendrorhous*. In *Methods in Biotechnology: Microbial Processes and Products*; Humana: Totowa, NJ, USA, 2005.
7. Gassel, S.; Breitenbach, J.; Sandmann, G. Genetic engineering of the complete carotenoid pathway towards enhanced astaxanthin formation in *Xanthophyllomyces dendrorhous* starting from a high-yield mutant. *Appl. Microbiol. Biotechnol.* **2014**, *98*, 345–350. [CrossRef] [PubMed]
8. Hara, K.Y.; Morita, T.; Mochizuki, M.; Yamamoto, K.; Ogino, C.; Araki, M.; Kondo, A. Development of a multi-gene expression system in *Xanthophyllomyces dendrorhous*. *Microb. Cell Factories* **2014**, *13*, 175. [CrossRef] [PubMed]
9. Sharma, R.; Gassel, S.; Steiger, S.; Xia, X.; Bauer, R.; Sandmann, G.; Thines, M. The genome of the basal agaricomycete *Xanthophyllomyces dendrorhous* provides insights into the organization of its acetyl-CoA derived pathways and the evolution of agaricomycotina. *BMC Genom.* **2015**, *16*, 233. [CrossRef] [PubMed]
10. Girard, P.; Falconnier, B.; Bricout, J.; Vladescu, B. β-Carotene producing mutants of *Phaffia rhodozyma*. *Appl. Microbiol. Biotechnol.* **1994**, *41*, 183–191. [CrossRef]
11. Ojima, K.; Breitenbach, J.; Visser, H.; Setoguchi, Y.; Tabata, K.; Hoshino, T.; van den Berg, J.; Sandmann, G. Cloning of the astaxanthin synthase gene from *Xanthophyllomyces dendrorhous* (*Phaffia rhodozyma*) and its assignment as a β-carotene 3-hydroxylase/4-ketolase. *Mol. Genet. Genom.* **2006**, *275*, 148–158. [CrossRef] [PubMed]
12. Pollmann, H.; Breitenbach, J.; Sandmann, G. Engineering of the carotenoid pathway in *Xanthophyllomyces dendrorhous* leading to the synthesis of zeaxanthin. *Appl. Microbiol. Biotechnol.* **2016**, *101*, 103–111. [CrossRef] [PubMed]
13. Albrecht, M.; Misawa, N.; Sandmann, G. Metabolic engineering of the terpenoid biosynthetic pathway of *Escherichia coli* for production of the carotenoids β-carotene and zeaxanthin. *Biotechnol. Lett.* **1999**, *21*, 791–795. [CrossRef]
14. Osawa, A.; Harada, H.; Choi, S.-K.; Misawa, N.; Shindo, K. Production of caloxanthin-β-D-glucoside, zeaxanthin 3,3-β-D-diglucoside, and nostoxanthin in a recombinant *Escherichia coli* expressing system harbouring seven carotenoid biosynthesis genes, including *crtX* and *crtG*. *Phytochemistry* **2011**, *72*, 711–716. [CrossRef] [PubMed]
15. Albrecht, M.; Takaichi, S.; Steiger, S.; Wang, Z.Y.; Sandmann, G. Novel hydroxycarotenoids with improved antioxidative properties produced by gene combination in *Escherichia coli*. *Nat. Biotechnol.* **2000**, *18*, 843–846. [CrossRef] [PubMed]
16. Nishida, Y.; Adachi, K.; Kasai, H.; Shizuri, Y.; Shindo, K.; Sawabe, A.; Komemushi, S.; Miki, W.; Misawa, N. Elucidation of a carotenoid biosynthesis gene cluster encoding a novel enzyme, 2,2′-β-hydroxylase, from *Brevundimonas*. sp. strain SD212 and combinatorial biosynthesis of new or rare xanthophylls. *Appl. Environ. Microbiol.* **2005**, *71*, 4286–4296. [CrossRef] [PubMed]
17. Borovkov, A.Y.; Rivkin, M.I. XcmI-Containing vector for direct cloning of PCR products. *BioTechniques* **1997**, *22*, 812–814.

18. Misawa, N.; Satomi, Y.; Kondo, K.; Yokoyama, A.; Kajiwara, S.; Saito, T.; Ohtani, T.; Miki, W. Structure and functional analysis of a marine bacterial carotenoid biosynthesis gene cluster and astaxanthin biosynthetic pathway proposed at the gene level. *J. Bacteriol.* **1995**, *177*, 6575–6584. [CrossRef] [PubMed]
19. Zhong, Y.J.; Huang, J.C.; Liu, J.; Li, Y.; Jiang, Y.; Xu, Z.F.; Sandmann, G.; Chen, F. Functional characterization of various algal carotenoid ketolases reveals that ketolating zeaxanthin efficiently is essential for high production of astaxanthin in transgenic *Arabidopsis*. *J. Exp. Bot.* **2011**, *62*, 3659–3669. [CrossRef] [PubMed]
20. Breitenbach, J.; Gerjets, T.; Sandmann, G. Catalytic properties and reaction mechanism of the CrtO carotenoid ketolase from the cyanobacterium *Synechocystis* sp. PCC 6803. *Arch. Biochem. Biophys.* **2013**, *529*, 86–91. [CrossRef] [PubMed]
21. Choi, S.K.; Matsuda, S.; Hoshino, T.; Peng, X.; Misawa, N. Characterization of bacterial β-carotene 3,3'-hydroxylases, CrtZ, and P450 in astaxanthin biosynthetic pathway and adonirubin production by gene combination in *Escherichia coli*. *Appl. Microbiol. Biotechnol.* **2006**, *72*, 1238–1246. [CrossRef] [PubMed]
22. Eugster, C.H. Chemical derivatization: Microsomale tests for the presence of common functional groups in carotenoids. In *Carotenoid Volume 1A Isolation and Analysis*; Britton, G., Liaaen-Jensen, S., Pfander, H., Eds.; Birkhäuser Verlag: Basel, Switzerland, 1995; pp. 71–80.
23. Steiger, S.; Perez-Fons, L.; Cutting, S.M.; Fraser, P.D.; Sandmann, G. Annotation and functional assignment of the genes for the C30 carotenoid pathways from the genomes of two bacteria *Bacillus indicus* and *Bacillus firmus*. *Microbiology* **2015**, *161*, 194–202. [CrossRef] [PubMed]
24. Schimming, O.; Fleischhacker, F.; Nollmann, F.I.; Bode, H.B. Yeast homologous recombination cloning leading to the novel peptides ambactin and xenolindicin. *ChemBioChem* **2014**, *15*, 1290–1294. [CrossRef] [PubMed]
25. Sandmann, G. Combinatorial biosynthesis of carotenoids in a heterologous host: A powerfull approach for the biosynthesis of novel structures. *ChemBioChem* **2002**, *3*, 629–635. [CrossRef]
26. Mata-Gómez, L.C.; Montañez, J.C.; Méndez-Zavala, A.; Aguilar, C.N. Biotechnological production of carotenoids by yeasts: an overview. *Microb. Cell Factories* **2014**, *13*, 12. [CrossRef] [PubMed]
27. Buchecker, R.; Liaaen-Jensen, S.; Borch, G.; Siegelman, H.W. Carotenoids of *Anacystis nidulans*, structures of caloxanthin and nostoxanthin. *Phytochemistry* **1976**, *15*, 1015–1018. [CrossRef]
28. Yokoyama, A.; Miki, W.; Zumida, H.; Shizuri, Y. New trihydroxy-keto-carotenoids Isolated from an astaxanthin-producing marine bacterium. *Biosci. Biotech. Biochem.* **1996**, *60*, 200–203. [CrossRef] [PubMed]
29. Beyer, P.; Kleinig, H.; Englert, G.; Meister, W.; Noaek, K. Carotenoids of Rhizobia. IV. Isolation and structure elucidation of the carotenoids of a mutant of *Rhizobium lupine*. *Helvetica Chim. Acta* **1979**, *260*, 2551–2557. [CrossRef]
30. Milanowska, J.; Gruszecki, W.I. Heat-induced and light-induced isomerization of the xanthophyll pigment zeaxanthin. *J. Photochem. Photobiol.* **2005**, *80*, 178–186. [CrossRef] [PubMed]
31. Rivera, S.M.; Christou, P.; Canela-Garayoa, R. Identification of carotenoids using mass spectrometry. *Mass Spectromet. Rev.* **2014**, *33*, 353–372. [CrossRef] [PubMed]
32. Ledetzky, N.; Osawa, A.; Iki, K.; Pollmann, H.; Gassel, S.; Breitenbach, J.; Shindo, K.; Sandmann, G. Multiple transformation with the *crtYB* gene of the limiting enzyme increased carotenoid synthesis and generated novel derivatives in *Xanthophyllomyces dendrorhous*. *Arch. Biochem. Biophys.* **2014**, *545*, 141–147. [CrossRef] [PubMed]

Journal of
Fungi

MDPI

Review

Carotenoid Biosynthesis in *Fusarium*

Javier Avalos [1,*], Javier Pardo-Medina [1], Obdulia Parra-Rivero [1], Macarena Ruger-Herreros [1], Roberto Rodríguez-Ortiz [1,2], Dámaso Hornero-Méndez [3] and María Carmen Limón [1]

[1] Departamento de Genética, Facultad de Biología, Universidad de Sevilla, 41012 Sevilla, Spain; jpardo6@us.es (J.P.-M.); duly@us.es (O.P.-R.); macarenarugerherreros@gmail.com (M.R.-H.); lrrodriguezor@conacyt.mx (R.R.-O.); carmenlimon@us.es (M.C.L.)
[2] Present Address: CONACYT-Instituto de Neurobiología-UNAM, Juriquilla, Querétaro 076230, Mexico
[3] Departamento de Fitoquímica de los Alimentos, Instituto de la Grasa, CSIC, Campus Universidad Pablo de Olavide, 41013 Sevilla, Spain; hornero@ig.csic.es
*   Correspondence: avalos@us.es; Tel.: +34-954-557-110

Received: 5 June 2017; Accepted: 4 July 2017; Published: 7 July 2017

Abstract: Many fungi of the genus *Fusarium* stand out for the complexity of their secondary metabolism. Individual species may differ in their metabolic capacities, but they usually share the ability to synthesize carotenoids, a family of hydrophobic terpenoid pigments widely distributed in nature. Early studies on carotenoid biosynthesis in *Fusarium aquaeductuum* have been recently extended in *Fusarium fujikuroi* and *Fusarium oxysporum*, well-known biotechnological and phytopathogenic models, respectively. The major *Fusarium* carotenoid is neurosporaxanthin, a carboxylic xanthophyll synthesized from geranylgeranyl pyrophosphate through the activity of four enzymes, encoded by the genes *carRA*, *carB*, *carT* and *carD*. These fungi produce also minor amounts of β-carotene, which may be cleaved by the CarX oxygenase to produce retinal, the rhodopsin's chromophore. The genes needed to produce retinal are organized in a gene cluster with a rhodopsin gene, while other carotenoid genes are not linked. In the investigated *Fusarium* species, the synthesis of carotenoids is induced by light through the transcriptional induction of the structural genes. In some species, deep-pigmented mutants with up-regulated expression of these genes are affected in the regulatory gene *carS*. The molecular mechanisms underlying the control by light and by the CarS protein are currently under investigation.

Keywords: neurosporaxanthin; xanthophyll; apocarotenoid; retinal; torulene; photoinduction; RING-Finger protein; carotenoid gene cluster

1. Introduction

The genus *Fusarium* is a complex group of phytopathogenic fungi, consisting of more than one thousand species [1], many of which have been the object of detailed attention [2]. Different *Fusarium* species were found to produce a large array of secondary metabolites. Some of them are pigments, and color variations have been used from long ago as distinctive traits of different strains (see, e.g., [3]). A most characteristic class of fungal pigments are the carotenoids, a family of lipophilic terpenoids ubiquitous in all major taxonomic groups [4,5]. The carotenoids are synthesized by all photosynthetic organisms, from cyanobacteria to higher plants, but they are also produced by diverse heterotrophic microorganisms, including fungi [6,7] and non-photosynthetic bacteria [8,9]. There are more than 750 natural carotenoids [5], providing typical yellowish, orange or reddish pigmentations to many plant organs, microorganisms and animals. With some notable exceptions in aphids [10], the animals are generally unable to synthesize carotenoids, and they get them through the diet. Carotenoids are especially relevant in photosynthetic species, where they play essential roles in light harvesting and photoprotection of the photosynthetic machinery [11] and in animals as a source for retinoids, i.e., retinal and retinoic acid [12]. However, the carotenoids do not seem to play essential roles in fungi, where their absence has no apparent phenotypic consequences apart of altered pigmentation [13].

As terpenoids, the synthesis of carotenoids derives from the condensation of C_5 isoprene units [4,5]. The universal terpenoid precursor is isopentenyl pyrophosphate (IPP), which can be synthesized either from mevalonate, produced from hydroxymethylglutaryl coenzyme A (HMG-CoA) (mevalonate pathway), or from derivatives of D-1-deoxyxylulose 1-phosphate, generated from the condensation of pyruvate and glyceraldehyde 3-phosphate (non-mevalonate pathway, [14]). In the cases investigated, fungal IPP is produced through the mevalonate pathway, while for carotenoids in bacteria and photosynthetic species IPP is produced through the non-mevalonate pathway, localized in plastids in the eukaryotic organisms. The early biosynthetic steps involve the sequential additions of IPP (isoprene, C_5) units to produce geranyl pyrophosphate (GPP, C_{10}), farnesyl pyrophosphate (FPP, C_{15}), and geranylgeranyl pyrophosphate (GGPP, C_{20}), carried out by prenyl transferases [4,5]. The first molecule with the characteristic aliphatic carotenoid-like structure is the colorless 15-*cis*-phytoene, consisting of a symmetrical polyene chain formed by the condensation of two GGPP units by phytoene synthase (Figure 1). The light-absorbing features of colored carotenoids are due to the presence of a chromophore, consisting of a series of conjugated double bonds generated by the activity of desaturase enzymes. Different subsequent chemical changes, usually the introduction of a cyclic end group (β and ε rings are the most common, while γ, κ, φ and χ rings are less frequent) on at least one of the ends of the molecule, and/or oxidative reactions (e.g., hydroxylation, epoxidation, carboxylation, esterification, etc), gives rise to the large diversity of known carotenoids [5].

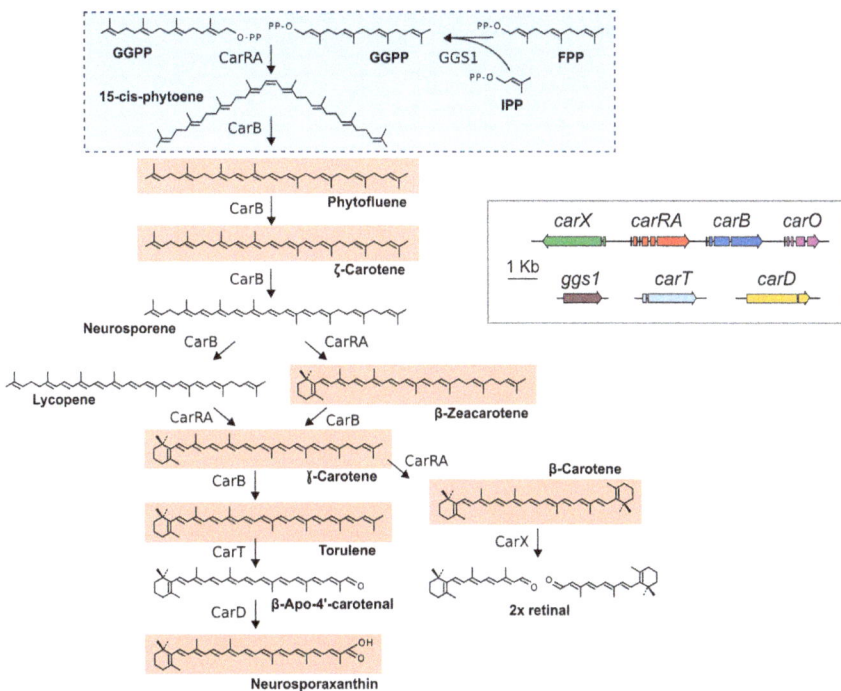

Figure 1. Carotenoid biosynthesis in *Fusarium*. Gene products responsible for the biosynthetic steps are indicated close to each arrow. Genomic organization of the genes is shown in the inserted box. Carotenoids detected in the High Performance Liquid Chromatography (HPLC) analyses described in Figure 4 are shaded in orange. Compounds shaded in blue (dotted box) are not detected in the chromatograms because of lack of absorption in the covered detection range (350–600 nm).

2. *Fusarium* Carotenoids

Early reports mentioned the occurrence of carotenoids in *Fusarium* species, e.g., *F. oxysporum* [3], but the first detailed biochemical analyses were described in *Fusarium aquaeductuum* [15]. Such analyses showed the occurrence of non-polar carotenoids and an acidic carotenoid fraction, which reminded the one assigned as the xanthophyll neurosporaxanthin (hereafter NX) in *Neurospora crassa* [16]. NX is a carboxylic C_{35} apocarotenoid, chemically defined as β-apo-4′-carotenoic acid [17]. NX is not a frequent carotenoid in fungi, and to date, besides the genera *Neurospora* and *Fusarium*, it has been only found in *Verticillium* [18,19] and *Podospora* [20].

Carotenoid analyses were refined in *F. aquaeductuum* in subsequent years, and confirmed the occurrence of NX together with precursor neutral carotenes, that included ζ-carotene, neurosporene, lycopene, γ-carotene, and torulene [21]. Later studies in *Fusarium fujikuroi*, and data presented in Section 5, corroborated the carotenoid composition of *F. aquaeductuum* with the exception of the detection of β-zeacarotene instead of lycopene. This indicates that the cyclization may be achieved either on neurosporene or on lycopene, depending on the species. Taken together, the chemical analyses are consistent with the biosynthetic pathway depicted in Figure 1. From phytoene, five desaturations and the cyclization of the non-saturated end yield the reddish torulene. The steps from torulene to NX were deduced from the enzymatic activities of the responsible enzymes, discussed below. In *F. fujikuroi* there are minor amounts of β-carotene [22], showing that the CarRA cyclase is also able to recognize γ-carotene as substrate. Because of its carboxylic group, NX may be a subject of esterification. In a marine *Fusarium* species, a fraction of the produced NX is accumulated as a glycosyl ester [23].

3. Genes and Enzymes of Carotenoid Metabolism in *Fusarium*

Some of the enzymes responsible for early steps of the terpenoid pathway were first identified in *F. fujikuroi* thanks to the similarity with formerly known enzymes. The typical cloning protocol consisted in the identification of conserved protein segments, the design of degenerate primers from such sequences to clone the corresponding internal gene segment by PCR, and the use of the amplified segment as a probe to clone the whole gene through the screening of a genomic library. Such methodology led to clone the genes for HMG-CoA reductase [24], producing mevalonate, and the prenyl transferases responsible for the synthesis of FPP [25] and its conversion to GGPP [26]. The mutants of these genes are expected to be non-viable because of their participation in the synthesis of essential terpenoids, and therefore their functions were inferred from their close relatedness with orthologs in other organisms. *F. fujikuroi* contains a second GGPP synthase located in the cluster for the synthesis of gibberellins, growth-inducing plant hormones of the terpenoid family. This second GGPP synthase gene is presumably specific for the synthesis of these secondary metabolites, as it indicates its coregulation with most of the genes from the gibberellin cluster [27].

The first gene specifically involved in the synthesis of carotenoids known in *Fusarium* was *carB* [28], identified by a PCR strategy similar to the one mentioned above through its sequence similarity to the *N. crassa* ortholog *al-1* [29]. The function of the gene was confirmed by complementation of a phytoene-accumulating albino mutant and by generation of the same phenotype by *carB* targeted mutation. The CarB enzyme is responsible for the five desaturation steps of the NX biosynthetic pathway, as it indicated the identification of a *carB* mutant allele specifically affected in the fifth desaturation [30]. This was in accordance with former observations with *al-1* of *N. crassa*: its expression in *Escherichia coli* or in vitro studies with purified AL-1 enzyme demonstrated its capacity to carry out the five desaturation steps for NX biosynthesis in this species [31]. Closely linked to *carB* it was found the gene *carRA* (Figure 1), orthologous of phytoene synthase *al-2* of *N. crassa* [32] and interpreted as the phytoene synthase of *Fusarium*, a conclusion consistent with the albino phenotype of their mutants [33]. In both species, a carboxylic domain with similarity to carotene cyclases precedes the phytoene synthase domain. The cyclase domain has a high conservation with that of the orthologous gene *crtYB* of the yeast *Xanthophyllomyces dendrorhous*, where the cyclase activity was biochemically demonstrated [34]. No mutants specifically affected in the cyclase activity have been described in

Fusarium, but the mutants of the cyclase domain of the *al-2* gene of *N. crassa* were shown to be defective in the cyclization step [35,36]. Taken together, the available data strongly suggest the same function for the amino domain of the CarRA protein of *F. fujikuroi*.

The reactions from torulene to NX include a carotenoid cleavage step, typically achieved by a family of enzymes known as carotenoid oxygenases [37]. A gene for an enzyme of this family, called *carX*, was found to be linked and transcribed divergently from the gene *carRA* (Figure 1). However, its targeted mutation did not impede the synthesis of NX [38], and the study of its enzymatic activity revealed that it cleaved β-carotene symmetrically to produce retinal [39]. Interestingly, the *carX/carRA/carB* cluster is linked to a rhodopsin gene, called *carO* because of its regulatory connections with the rest of the genes of the cluster (see next sections). The rhodopsins typically bind retinal as a light-absorbing prosthetic group, providing coherence to the gene organization of the cluster: the genes *carRA*, *carB*, and *carX* are necessary to produce retinal, which is presumably used by the rhodopsin encoded by the coregulated *carO* gene. Retinal might be also used for a second rhodopsin, encoded by the gene *opsA* [40]. In addition, retinal might be subject to further chemical reactions. The *Fusarium* genomes contain a gene for an aldehyde dehydrogenase highly similar to RALDH enzymes converting retinal to retinoic acid in mammals. This enzyme, that was called CarY, was investigated in *F. verticillioides* and found to exhibit such enzymatic activity in vitro [41]. Targeted mutation of the *carY* gene in *F. verticillioides* had no effect on carotenogenesis, but resulted in diverse developmental alterations. However, no retinoic acid could be detected in the control fungal cells and hence the biochemical function of the CarY enzyme in *Fusarium* has not been solidly established.

The *Fusarium* genes responsible for the conversion of torulene to NX have been recently described. The search for genes encoding other putative carotenoid oxygenases in the *Fusarium* genome databases led to identify the gene *carT*, whose function was revealed by the presence of a mutation in a torulene accumulating mutant and its genetic complementation with the wild-type *carT* allele [42]. As biochemical confirmation, purified CarT enzyme efficiently cleaved torulene in vitro to produce β-apo-4'-carotenal. As additional support, the targeted mutation of its ortholog in *N. crassa*, called *cao-2* from carotenoid oxygenase, resulted in the block of NX production and the accumulation of torulene [43]. These findings established a new enzymatic class in the carotenoid oxygenase family. The function of the gene *carT* in *Fusarium* carotenogenesis, together with those of the genes *carRA* and *carB*, was also confirmed by targeted gene disruption in *Gibberella zeae*, teleomorph of *Fusarium graminearum* [44].

The product of CarT, β-apo-4'-carotenal, requires a further oxidation to produce NX. The responsible gene was first discovered in *N. crassa* thanks to the study of a yellow mutant, called *ylo-1*. This mutant exhibited a puzzling biochemical phenotype, since it contained a complex carotenoid mixture that varied with the culture conditions and that did not include NX [45,46]. The gene *ylo-1* encodes an aldehyde dehydrogenase mediating the last step of NX biosynthesis, as demonstrated in its ability to complement the *ylo-1* mutation [47] and the capacity of purified YLO-1 protein to convert in vitro β-apo-4'-carotenal into NX. This finding allowed the identification of the orthologous gene in *F. fujikuroi*, that was called *carD* [48]. The function of *carD* was confirmed by targeted mutation, which resulted in the lack of NX and the accumulation of unusual carotenoids, interpreted as β-apo-4'-carotenal derivatives.

4. Regulation by Light

Early observations in *F. oxysporum* revealed that illumination promotes the accumulation of carotenoids in the mycelium [3]. First detailed studies on the effect of light were carried out with *F. aquaeductuum*. In an initial report, focused on the effect of temperature on the photoinduction process [49], the time course of the response to a transient light exposure revealed a gradual carotenoid accumulation that reached a maximum at about 12 h after illumination. However, the synthesis keeps increasing at a slower rate for at least three days if the culture is maintained under light [50].

Regarding the effect of temperature, the secondary reaction (formation of colored carotenoids) decreases at lower temperatures while the primary reaction to light, presumably a photochemical process, is independent of temperature in the range of 5–25 °C.

The photoinduction is ineffective under anaerobic conditions, as indicated by the lack of carotenoid accumulation if the illuminated mycelia are transferred to an oxygen-free atmosphere [51]. However, the photoinduced state is maintained, as indicated by the onset of carotenoid biosynthesis if the aerobic conditions are restored, even in the dark. Under aeration, the photoinduction requires protein synthesis, as shown by the lack of response if cycloheximide is added before or immediately after light exposure [50]. As happens with anaerobiosis, the removal of cycloheximide allows the start of carotenoid accumulation at any time in the dark, at least up to 30 h, pointing again to a high stability of the induced state of the photoreception system. In an ingenious experiment, when a *F. aquaeductuum* culture was illuminated, incubated in the dark to allow the start of enzyme production, and then exposed to anaerobiosis and to cycloheximide to block further enzyme production, there was no carotenoid production. However, if the culture was returned afterwards to aerobiosis, carotenoid biosynthesis started even in the presence of cycloheximide, indicating that the effect of anaerobiosis is at the level of enzyme activity, and that the enzymes are sufficiently stable [51].

In *F. aquaeductuum*, the effect of light may be partially replaced by the addition of the sulfhydryl oxidizing reagents p-chloro- and p-hydroxymercuribenzoate [50,52] or the oxidative reagent hydrogen peroxide [53], suggesting that oxidation of -SH groups plays a role in the light detection system. Accordingly, the photoinduction is abolished upon addition of reducing agents as dithionite and hydroxylamine, but the response is recuperated if the reducing agent is removed, indicating the recovery of the photoreceptor system [53]. However, while a short exposure to light is sufficient for a sustained photoinduction, p-hydroxymercuribenzoate must be all the time present to maintain its stimulating effect, and such stimulation is additive with that of light [54], indicating different mechanisms of action. Moreover, p-hydroxymercuribenzoate was ineffective in *F. fujikuroi* in the dark, while this species exhibits a similar photoinduction [55].

The study of light dependence of the photoinduction in *F. aquaeductuum* [56] showed that the amount of produced carotenoids depends on the incident light over a 100-fold range, with a reciprocity law holding true over a wide range of exposure times and light intensities. Action spectrum for NX photoinduction extends from 400 to 500 nm, with maxima at 375/380 nm and 450/455 nm and a shoulder at 430/440 nm, a shape consistent with the participation of a flavin photoreceptor. No induction is detected with wavelengths beyond 500 nm, and actually red light proved to be ineffective, discarding the participation of a phytochrome-like photoreceptor [57]. However, incubation of the fungus with methylene blue or toluidine blue allows it to respond to red light [58], suggesting that these chemicals may act as artificial photoreceptors. The shape of the action spectrum also discards the carotenoids as light-absorbing chromophores. As supporting evidence, the accumulation of phytoene in the albino *carB* mutant SG43 is still dependent on light [22], and the mutants of gene *carX* [38], involved in retinal formation, or those of the rhodopsin genes *carO* [59] and *opsA* [40] exhibit a full carotenoid photoinduction.

In *N. crassa*, a fungus with a similar action spectrum for light-induced carotenoid biosynthesis [60], the photoinduction is totally dependent on the White Collar complex, consisting of the photoreceptor WC-1 and its partner WC-2 (reviewed by [61]). However, the mutants of the *wc-1* orthologous genes of *F. fujikuroi* (*wcoA* [62]) and *F. oxysporum* (*wc1* [63]) conserve to different extents a detectable carotenoid photoinduction under continuous illumination. The targeted mutation of the orthologous *wc-1* and *wc-2* genes in *F. graminearum*, *fgwc-1* and *fgwc-2*, result in a paler pigmentation of the surface colonies under light, but the levels of carotenoids either in the wild type and the mutants were not chemically determined [64].

The analysis of carotenoid accumulation on surface cultures revealed a two-stage response in *F. fujikuroi*, with a first rapid increase dependent on WcoA and a slower subsequent carotenoid accumulation depending on another flavin photoreceptor, the DASH cryptochrome CryD [65].

The photoactivity of this flavin photoreceptor has been experimentally demonstrated [66], and its participation as a second photoreceptor explains the maintenance of photoinduced carotenoid accumulation in the *wcoA* mutants under constant illumination. The photoreceptors WcoA and CryD are not only involved in the regulation of carotenogenesis, as indicated by the alteration in the production of other pigments and secondary metabolites in their corresponding mutants [62,67], even in the dark in the case of WcoA.

The photoinduction of *Fusarium* carotenogenesis results from a rapid increase in the transcript levels of most of the structural genes. *Northern* blot experiments in *F. fujikuroi* showed similar induction kinetics for the four genes of the *car* cluster, *carRA, carB, carO* [59], and *carX* [38], as well as *carT* [42]. The transcripts reach maximal levels after one hour of exposure to light and decay afterwards, a down-regulating phenomenon known as photoadaptation. This photoinduction pattern has been confirmed by RT-qPCR approaches (Figure 2), and similar results have been obtained in *F. oxysporum* [68] and *F. verticillioides* [69]. The transcriptional photoresponse is similar to the one exhibited by the orthologous genes in *N. crassa* (reviewed by [70]), with the exception of the GGPP synthase gene *al-3*, that exhibits in this fungus a strong photoinduction while its *F. fujikuroi* ortholog *ggs1* is hardly affected by light [26]. Our recent RNA-seq data have revealed however a significant photoinduction of *ggs1* mRNA (Figure 3). A minor photoresponse was exhibited by the gene *carD* from *F. fujikuroi* [47], corroborated by the RNA-seq data (Figure 3), while no photoinduction was detected in the case of *ylo-1* from *N. crassa* [48], both genes encoding the ALDH enzymes responsible for the last reaction for NX production (Section 3). Therefore, in *F. fujikuroi*, the whole NX biosynthetic pathway is regulated by light.

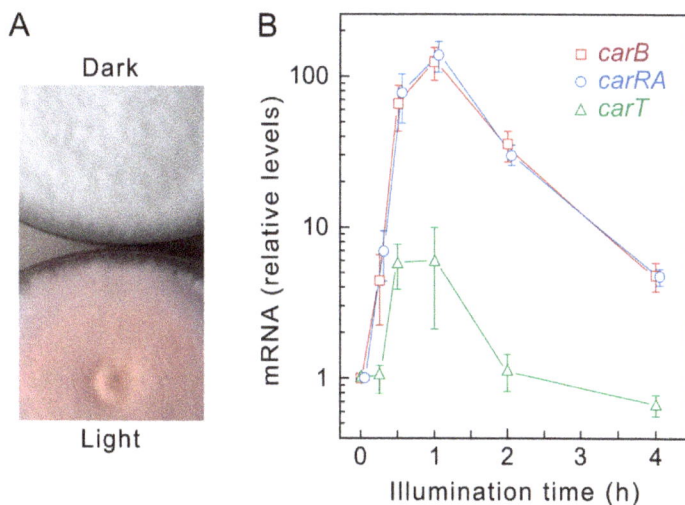

Figure 2. Photoinduction of carotenogenesis in *F. fujikuroi*. (**A**) Aspect of 7-day old colonies of the wild type IMI58289 grown on minimal medium in the dark or under continuous illumination; (**B**) Example of the kinetics of mRNA accumulation for the genes *carRA, carB*, and *carT* in the wild type FKMC1995. Transcript levels were determined by RT-qPCR (real time quantitative PCR) and referred to those of β-tubulin gene. Levels in the dark for each gene were taken as 1 (adapted from [65]).

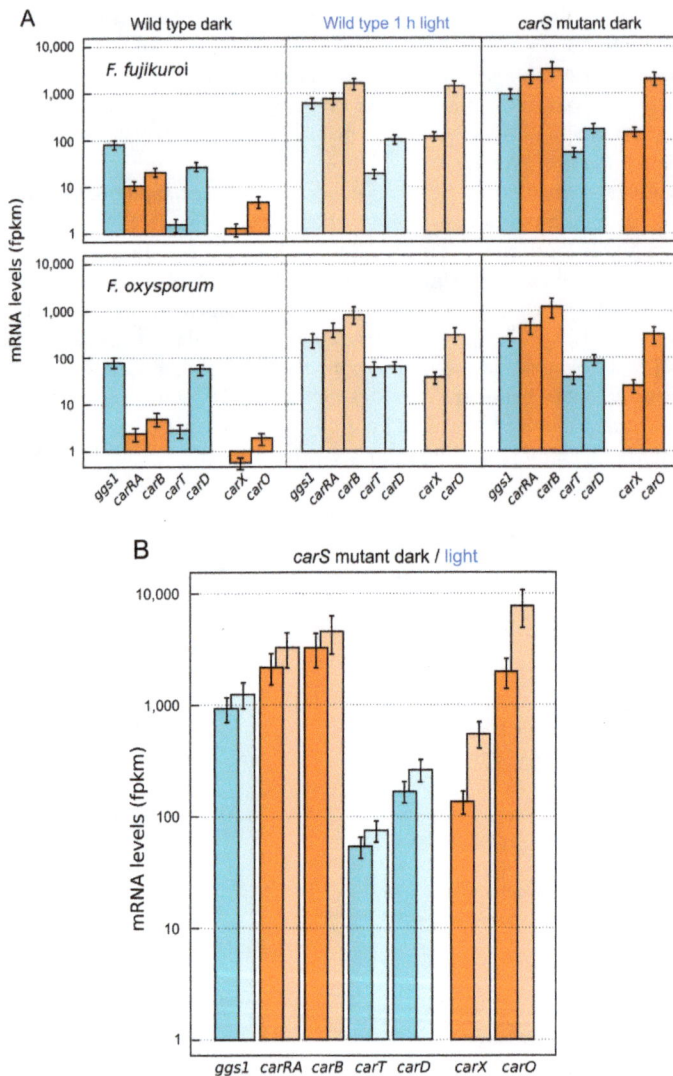

Figure 3. Effect of light and *carS* mutation on the expression levels for the *Fusarium* genes involved in carotenoid metabolism. The genes are grouped according to their functions: *ggs1*, *carRA*, *carB*, *carT*, and *carD* involved in NX biosynthesis and *carX* and *carO* involved in retinal and CarO rhodopsin production. Orange colors correspond to the genes organized as a cluster, and blue color to genes unlinked in the genome. Darker colors indicate cultures incubated in the dark and paler colors indicate cultures illuminated for one hour. (**A**) Above: wild-type strain IMI58289 of *F. fujikuroi* and its *carS* mutant SG39. Below: wild type strain 4287 of *F. oxysporum* (f. sp. *lycopersici*) and its *carS* mutant SX2; (**B**) Effect of light on the mRNA levels of the *carS* mutant SG39. RNA-seq data and culture conditions correspond to analyses already described [71]. In brief, all strains were cultured for three days in dark in minimal medium (DG medium for *F. fujikuroi* and DG with 3 g asparagine instead of NaNO$_3$ for *F. oxysporum*). In the cases indicated, the cultures were exposed to one hour of illumination. RNA samples were sequenced with Illumina technology and the resulting data were analyzed with software tools for read mapping (Bowtie and TopHat), transcriptome assembly (Cuffmerge and Cufflinks), and differential gene expression analysis (Cuffdiff and CummeRbund).

The study of the transcriptional photoinduction of the structural genes for carotenogenesis in the mutants of the *wcoA* and *cryD* genes is consistent with different mechanisms of action for the encoded photoreceptors [65]. After illumination, the *wcoA* mutants exhibit photoinduction of carotenogenesis, but they accumulate carotenoids more slowly and their total levels do not reach those of the wild type. However, both the carotenoid content and the mRNA levels for the structural *car* genes are much lower in the dark in the *wcoA* mutants than in the wild type. Even more, the transcriptional photoinduction of the *car* genes is basically absent in the *wcoA* mutants, with mRNA levels after illumination lower than those of the wild type in the dark. Therefore, the induction of carotenogenesis by the alternative photoreceptor, presumably CryD, must be done through a transcription-independent mechanism. The molecular basis for such a mechanism, that could involve mRNA stability or translation and enzymatic activities or stabilities, remains to be elucidated.

The decay of the mRNA levels after photoinduction has been investigated in detail in *N. crassa*. In this fungus, the small flavin prohotoreceptor VIVID (VVD), its name due to the higher carotenoid levels of the *vvd* mutants under light, plays a key role in deactivating the WC photoreceptor system after illumination [70]. In contrast, the mutation of the VIVID ortholog in *F. fujikuroi*, called *vvdA*, results in a lower carotenoid accumulation in this fungus [72]. The kinetics of carotenogenesis in the *vvdA* mutants show a faster carotenoid accumulation immediately after light exposure, consistent with an attenuating function of VvdA on the photoactivated WcoA, but a slower accumulation after more prolonged growth under light, suggesting an up-regulating role during the second stage of carotenoid photoinduction [65].

5. Down-Regulation of Carotenoid Production: The *CarS* Gene

Standard mutagenesis protocols allow the identification of *Fusarium* mutants affected in the synthesis of carotenoids, easy to detect because of their changes in pigmentation. Such mutants are more easily identified on minimal medium, such as the DG minimal agar [73], than on richer media, e.g., Potato Dextrose Agar (PDA), which frequently results in the accumulation of other unrelated pigments which may mask the color of carotenoids. Carotenoid mutants usually identified by visual inspection are either albino in the light, as those lacking an early enzymatic step of the carotenoid pathway, or deeply pigmented in the dark, affected in a down-regulating function. The deep-orange mutants, exhibiting a light-independent carotenoid production and generically called *carS*, have been described in *F. fujikuroi* [22] and *F. oxysporum* [68]. The *carS* mutants have been investigated in detail in *F. fujikuroi*, where they accumulate an increasing amount of carotenoids with aging [55] and contain large amounts of mRNA for the structural genes, from *carRA* to *carD* [38,42,48,59] (Figure 3). This up-regulation results in an increase of carotenogenic enzymes, as deduced from the study of in vitro carotenogenesis in a cell-free system of a *carS* mutant compared to the wild type [74]. Because of their high carotenoid content, the effect of light is less apparent in the *carS* mutants, but expression studies have found a detectable photoinduction of the transcript levels for the structural genes. This has been found in *F. fujikuroi* [42] and in *F. oxysporum* [68], and it has been recently corroborated in *F. fujikuroi* in RNA-seq analyses (Figure 3).

The qualitative mixture of carotenoids accumulated by two *carS* mutants of *F. fujikuroi* obtained in different genetic backgrounds is similar to that of their corresponding wild types under light, with minor differences, mainly a lower torulene content in the *carS* strains (Figure 4). Because of the large accumulation of carotenoids in the dark, these mutants are also useful to investigate possible damaging effects of light on the carotenoids. The comparison of the chromatograms from dark or light-grown cultures in two *carS* mutants reveal again only minor changes, with a decrease in torulene and an increase in β-carotene in the samples from the illuminated cultures (Figure 4), which could be due to different sensitivities of these carotenes to light. Because of their intense pigmentation, the *carS* mutants have been useful tools to detect mutants with qualitative alterations in the carotenoid mixtures, e.g., making possible the identification of torulene accumulating mutants [22] or mutants with alterations in substrate recognition by the CarB desaturase [30]. The *carS* mutants have been also an optimal background to check the effect of possible carotenoid inhibitors on *Fusarium* carotenogenesis, revealing the differential efficiency of some cyclase inhibitors that were fully active in other species [55].

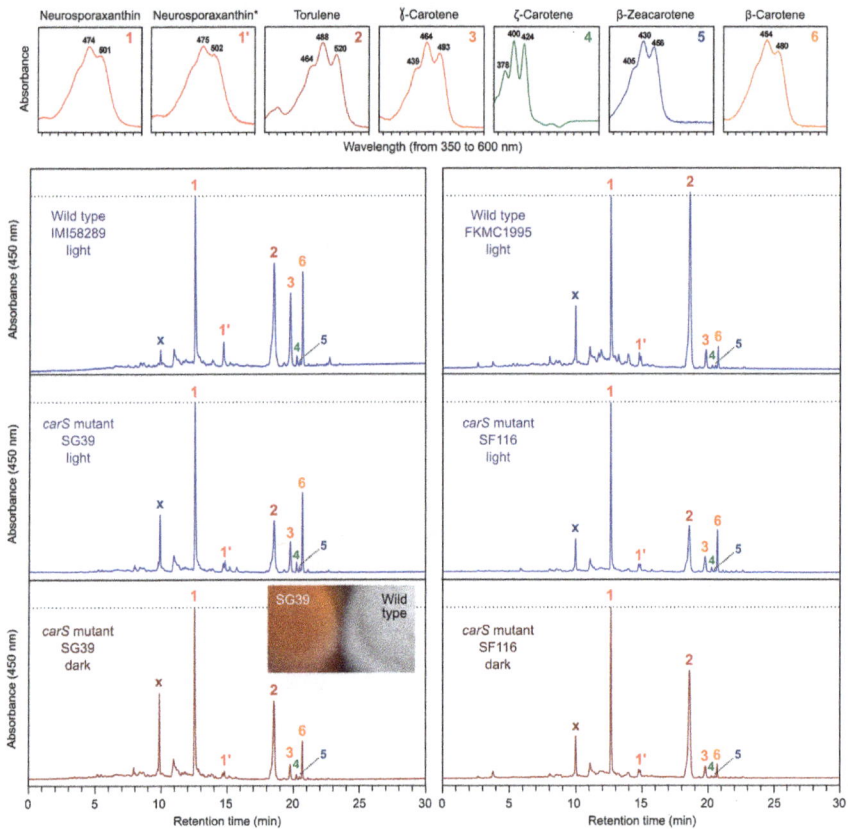

Figure 4. HPLC analysis of carotenoids produced by two different wild types of *F. fujikuroi* and the *carS* mutants SG39 [22] and SF116 [48]. The strains were grown for 7 days on DG minimal medium in the dark or under light (7 W m^{-2} white light). The method for HPLC separation has been described [75]. In brief, the samples were run in a reverse-phase C18 column with a binary-gradient elution (acetone : deionized water) at a flow rate of 1 mL/min. Detection was performed at 450 nm and online UV/Visible absorption spectra were acquired in the wavelength range 350–600 nm. The UV/Visible spectra of the identified carotenoids are shown above. The peak X corresponds to an unknown carotenoid with maximal absorbance at 480 nm, whose identity is under investigation. The peak 1' is assigned as a neurosporaxanthin isomer. Phytofluene cannot be detected at 450 nm, but traces of this carotene were detected at 368 nm (data not shown). The inner picture shows the pigmentation of 7-day old colonies of the wild strain IMI58289 and the *carS* mutant SG39 grown on minimal DG medium in the dark.

The gene *carS* was found to code for a protein of the RING finger (RF) family. The assignation was based in the identification of relevant mutations in all the *carS* mutants analyzed so far, the deep-pigmented phenotype resulting from the targeted mutation of the gene in *F. oxysporum* [68] and the complementation of the *carS* mutation in *F. fujikuroi* [76]. The predicted CarS protein has sequence similarity with the protein CrgA of *M. circinelloides*, whose mutation results in a similar carotenoid overproducing pattern in this species [77,78]. The degree of similarity between the CarS and CrgA proteins is not very high, as expected for two taxonomically distant fungi, but it covers the most relevant CrgA domains. These include two amino-terminal RF domains, one of them initially disregarded in CarS because of a wrong intron assignation in the *F. fujikuroi* genome annotation [71], and a LON protease domain [79]. However, CrgA has two glutamine-rich regions and a carboxy-terminal isoprenylation site, the first one underrepresented and the second absent in CarS. In other proteins, the RF domains interact with E3 ligase-type enzymes that mediate ubiquitylation of target proteins, frequently as a label for their degradation. The function of CrgA of *M. circinelloides* has received considerable attention, and the available information may provide some clues on the hypothetical way of action of CarS in *Fusarium*. At least one of the RF domains of CrgA is essential for its regulatory function in carotenogenesis, suggesting that this protein might function as an E3 ubiquitin ligase [78]. However, CrgA interacts with one of the three WC proteins of *M. circinelloides*, MCWC-1b, to trigger its degradation through an ubiquitylation-independent mechanism [80]. There are other similarities between CarS and CrgA functions. First, as found for the *carS* mutants in *Fusarium*, carotenoid biosynthesis maintains a detectable photoinduction in the *crgA* mutants of *M. circinelloides* [77]. Second, both CrgA and CarS are involved in other processes in addition to carotenogenesis, pointing to wider regulatory functions. This is supported by the alterations in growth and sporulation by the *crgA* mutants of *M. circinelloides* [81,82], and in gibberellin and bikaverin productions by the *carS* mutants of *F. fujikuroi* [83,84].

Interestingly, *carS* mutants have not been described in other *Fusarium* species and they could not be found in mutagenesis screenings in *F. verticillioides* (J. García-Martínez and V. Díaz-Sánchez, unpublished observations). However, the *carS* gene of *F. fujikuroi* is able to restore a basal carotenoid production upon introduction in a *carS* mutant of *F. oxysporum*, although not totally at the level of expression of the structural genes, indicating a functional conservation in both species [68].

6. Regulation by Light-Independent Factors

Nitrogen availability controls the production of different *Fusarium* secondary metabolites. Well-known examples are the productions of gibberellins, bikaverin, and fusarubin, induced by the absence of nitrogen through a complex regulatory network [85]. However, other metabolites, as fusarins and fusaric acid, exhibit an opposite regulatory pattern, with a higher production under excess of nitrogen [86]. Experiments with immobilized mycelia of *F. fujikuroi* incubated under low nitrogen conditions, to which excess nitrogen was subsequently added, revealed a higher synthesis of carotenoids under nitrogen starvation [87]. The negative effect of nitrogen was confirmed with the exchange of media between immobilized mycelia samples with different nitrogen contents, resulting in enhanced or reduced synthesis depending on the pass to low or high nitrogen conditions, respectively, indicating that the repressing effect is reversible.

In a different approach, the effect of nitrogen was studied in shake cultures of wild type and *carS* mutants. The production of carotenoids was higher in low nitrogen medium (low N/C ratio) than in high nitrogen medium (high N/C) in either of the tested strains [84]. A different set of experiments, in which cultures grown under an excess of nitrogen were transferred to a nitrogen-free solution, showed an increase of transcript levels of the structural genes *carRA* and *carB* after the transfer, which was accompanied by an enhanced accumulation of carotenoids. As found for photoinduction (see Section 3), the transcriptional increase was transitory, with maximal levels found in this case after more than 10 h following nitrogen deprivation.

The stimulatory effect of nitrogen starvation on *F. fujikuroi* carotenogenesis, either at mRNA or carotenoid levels, is additive with the one produced by light [84], indicating different activating mechanisms. The regulation of carotenogenesis by nitrogen has been also described in *N. crassa*, where the mRNA levels for the *carRA* and *carB* orthologs *al-1* and *al-2* increase considerably under nitrogen limitation compared to standard nitrogen conditions, either in the wild type or in blind *wc* mutants [88]. The regulation by nitrogen might involve control of expression at the level of chromatin structure, as indicated by the differences in histone methylation found for the genes of the *car* cluster in a mutant of the methyltransferase KMT6 in *Fusarium graminearum* [89]. In *F. fujikuroi*, the expression of several nitrogen-regulated clusters for secondary metabolite production correlate with H3K9 acetylation [90]. This suggests a similar mechanism in the regulation of the structural *car* genes, which is currently under investigation.

Other regulatory circuits may also influence the control of carotenoid biosynthesis. A major regulatory pathway is the one involving cAMP, produced by adenylate cyclase under the stimulation of a Gα unit from a heterotrimeric G protein. cAMP acts on other proteins, including protein kinases that phosphorylate target proteins to module their activity. The mutation of the adenylate cyclase gene *acyA* in *F. fujikuroi* results in higher levels of carotenoids in the dark but in a reduced photoinduction [91]. The mutation affects other phenotypes, as the growth pattern or the production of other secondary metabolites. Another study in the same species, that extended the analysis to the effect of mutations of two adenylyl cyclase-stimulating Gα subunits and two cAMP-dependent PKs, confirmed different alterations in secondary metabolism and development [92], but in this case the effect on carotenogenesis was not investigated. Carotenogenesis may also have a regulatory connection with sexual development, as suggested by the lower carotenoid photoinduction of the mutants of the MAT1-2-1 mating-type gene of *F. verticillioides* [69].

In *N. crassa*, besides the regulation by light, the synthesis of carotenoids is coupled to conidiation in a light-independent manner [70]. Actually, conidiation is stimulated by light and aerial growth in this fungus, leading to a massive conidia production that provides a typical orange pigmentation in slant cultures. Conidiation is less abundant in *Fusarium*, and the synthesis of carotenoids in the conidia has been usually disregarded. A comparison of the effect of light on the amount of carotenoids in mycelia and conidia from surface cultures in a wild type and a *carS* mutant of *F. oxysporum* showed the conservation in the conidia of the regulation by light and the deregulation by the *carS* mutation [93] (Figure 5). The low amount of carotenoids in wild type conidia in the dark indicates the lack of a developmental induction as the one described in *N. crassa*.

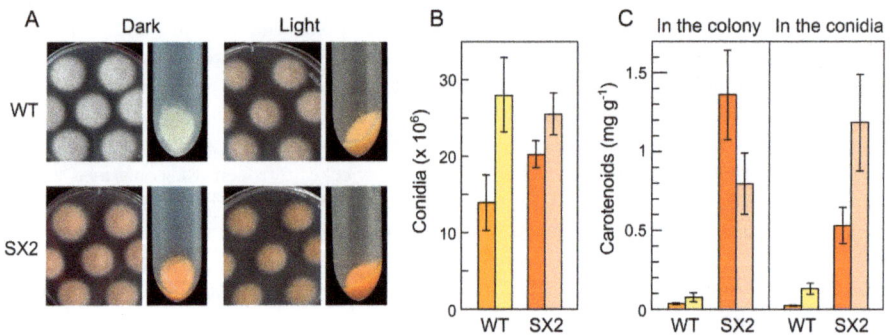

Figure 5. Conidiation and carotenoid biosynthesis in *F. oxysporum*. (**A**) Aspect of colonies of the wild type strain 4287 of *F. oxysporum* (f. sp. *lycopersici*) and its *carS* mutant SX2 grown on DG minimal medium at 30 °C for 7 days in the dark or under 5 w m^{-2} white light. Pictures on the right show the aspect of precipitated conidia obtained from the colonies as described [59]; (**B**) total amounts of conidia per Petri dish produced by the same strains under the indicated culture conditions; (**C**) total amounts of carotenoids accumulated in the mycelia and in the conidia mentioned above. The strains were grown in the dark (left darker bars) or under light (right paler bars). Data adapted from [93].

The cellular location of carotenoid biosynthesis is a regulatory aspect that has received little attention. Because of their hydrophobic nature, the carotenoids are assumed to interact with membranes, but their subcellular distribution in *Fusarium* is unknown. A biochemical approach based on the specific labeling of different terpenoids from ^{14}C-labeled mevalonate allowed to establish that the carotenoids are produced in different cell compartments than those where the gibberellins and the sterols are synthesized [94]. Experiments to visualize the physical location of the enzymes for carotenoid biosynthesis in the fungal cells, based in their fusion to fluorescent proteins, are currently underway.

7. Conclusions and Future Prospects

Following the early studies in *F. aquaeductuum*, the fungus *F. fujikuroi*, and more recently also *F. oxysporum*, have become reference models in the research of fungal carotenogenesis and a wealth of information has been accumulated on the biochemistry and genetics of their NX production. All the genes and enzymes of the biosynthetic pathway are known and current research work is dedicated to elucidate the molecular mechanisms of the regulation. The available information on the control by light has revealed an unexpected complexity in relation to the model system *N. crassa*, in which the White Collar complex is the only transcriptional activator of the structural genes for neurosporaxanthin biosynthesis. At least a second photoreceptor, the DASH cryptochrome CryD, is also involved in the photoresponse in *Fusarium*. Our current research efforts are centered in a better understanding of the regulation by light and by the CarS protein, and their possible connections. Studies on the mechanism of action of CarS are focused in the alteration of its expression and in the identification of interacting proteins. Efforts are also centered in the identification of regulatory proteins able to bind the promoters of the structural *car* genes, which might include CarS targets. The regulation may include epigenetic mechanisms, as those related with chromatin structure, also under study. While the biochemistry of the carotenoid pathway has many similarities between different producing fungi, the regulatory mechanisms have been the object of a higher diversification, allowing refined adaptations to respond more efficiently to the different needs in their respective natural habitats. The studies on the regulatory processes governing carotenoid biosynthesis in *Fusarium* are a promising area of research, which predictably may lead to the discovery of novel molecular mechanisms.

Acknowledgments: We thank the Spanish Government (projects BIO2012-39716, BIO2015-69613-R, AGL2014-53195R and CaRed Network BIO2015-71703-REDT) and Junta de Andalucía (project CTS-6638) for financial support.

Conflicts of Interest: The authors declare no conflict of interest.

References

1. Booth, C. *The Genus Fusarium*; Commonwealth Agricultural Bureaux for the Commonwealth Mycological Institute: Farnham Royal, UK, 1971.
2. Leslie, J.F.; Summerell, B.A. *The Fusarium Laboratory Manual*; Blackwell Professional: Ames, IA, USA, 2006.
3. Carlile, M.J. A study of the factors influencing non-genetic variation in a strain of *Fusarium oxysporum*. *J. Gen. Microbiol.* **1956**, *14*, 643–654. [CrossRef] [PubMed]
4. Britton, G.; Liaaen-Jensen, S.; Pfander, H. *Carotenoids*; Birkhäuser Verlag: Basel, Switzerland, 1998; Volumes 1 and 2.
5. Britton, G.; Liaaen-Jensen, S.; Pfander, H. *Carotenoids: Handbook*; Birkhauser Verlag: Basel, Switzerland, 2004.
6. Sandmann, G.; Misawa, N. Fungal carotenoids. In *The Mycota X. Industrial Applications*; Osiewacz, H.D., Ed.; Springer: Berlin, Germany, 2002; pp. 247–262.
7. Avalos, J.; Díaz-Sánchez, V.; García-Martínez, J.; Castrillo, M.; Ruger-Herreros, M.; Limón, M.C. Carotenoids. In *Biosynthesis and Molecular Genetics of Fungal Secondary Metabolites*; Martín, J.F., García-Estrada, C., Zeilinger, S., Eds.; Springer: New York, NY, USA, 2014; pp. 149–185.
8. Phadwal, K. Carotenoid biosynthetic pathway: Molecular phylogenies and evolutionary behavior of *crt* genes in eubacteria. *Gene* **2005**, *345*, 35–43. [CrossRef] [PubMed]
9. Boronat, A.; Rodríguez-Concepción, M. Terpenoid biosynthesis in prokaryotes. *Adv. Biochem. Eng. Biotechnol.* **2015**, *148*, 3–18. [PubMed]
10. Mandrioli, M.; Rivi, V.; Nardelli, A.; Manicardi, G.C. Genomic and cytogenetic localization of the carotenoid genes in the aphid genome. *Cytogenet. Genome Res.* **2016**, *149*, 207–217. [CrossRef] [PubMed]
11. Domonkos, I.; Kis, M.; Gombos, Z.; Ughy, B. Carotenoids, versatile components of oxygenic photosynthesis. *Prog. Lipid Res.* **2013**, *52*, 539–561. [CrossRef] [PubMed]

12. Blomhoff, R.; Blomhoff, H.K. Overview of retinoid metabolism and function. *J. Neurobiol.* **2006**, *66*, 606–630. [CrossRef] [PubMed]
13. Avalos, J.; Limón, M.C. Biological roles of fungal carotenoids. *Curr. Genet.* **2015**, *61*, 309–324. [CrossRef] [PubMed]
14. Rohdich, F.; Kis, K.; Bacher, A.; Eisenreich, W. The non-mevalonate pathway of isoprenoids: Genes, enzymes and intermediates. *Curr. Opin. Chem. Biol.* **2001**, *5*, 535–540. [CrossRef]
15. Rau, W.; Zehender, C. Die Carotinoide von *Fusarium aquaeductuum* Lagh. *Arch. Mikrobiol.* **1959**, *32*, 423–428. [CrossRef] [PubMed]
16. Zalokar, M. Isolation of an acidic pigment in *Neurospora. Arch. Biochem. Biophys.* **1957**, *70*, 568–571. [CrossRef]
17. Aasen, A.J.; Jensen, S.L. Fungal carotenoids II. The structure of the carotenoid acid neurosporaxanthin. *Acta Chem. Scand.* **1965**, *19*, 1843–1853. [CrossRef] [PubMed]
18. Valadon, L.R.G.; Mummery, R.S. Biosynthesis of neurosporaxanthin. *Microbios* **1969**, *1A*, 3–8.
19. Valadon, L.R.G.; Osman, M.; Mummery, R.S.; Jerebzoff-Quintin, S.; Jerebzoff, S. The effect of monochromatic radiation in the range 350 to 750 nm on the carotenogenesis in *Verticillium agaricinum. Physiol. Plant.* **1982**, *56*, 199–203. [CrossRef]
20. Strobel, I.; Breitenbach, J.; Scheckhuber, C.Q.; Osiewacz, H.D.; Sandmann, G. Carotenoids and carotenogenic genes in *Podospora anserina*: Engineering of the carotenoid composition extends the life span of the mycelium. *Curr. Genet.* **2009**, *55*, 175–184. [CrossRef] [PubMed]
21. Bindl, E.; Lang, W.; Rau, W. Untersuchungen über die lichtabhängige Carotinoidsynthese. VI. Zeitlicher Verlauf der Synthese der einzelnen Carotinoide bei *Fusarium aquaeductuum* unter verschiedenen Induktionsbedingungen. *Planta* **1970**, *94*, 156–174. [CrossRef] [PubMed]
22. Avalos, J.; Cerdá-Olmedo, E. Carotenoid mutants of *Gibberella fujikuroi. Curr. Genet.* **1987**, *25*, 1837–1841. [CrossRef]
23. Sakaki, H.; Kaneno, H.; Sumiya, Y.; Tsushima, M.; Miki, W.; Kishimoto, N.; Fujita, T.; Matsumoto, S.; Komemushi, S.; Sawabe, A. A new carotenoid glycosyl ester isolated from a marine microorganism, *Fusarium* strain T-1. *J. Nat. Prod.* **2002**, *65*, 1683–1684. [CrossRef] [PubMed]
24. Woitek, S.; Unkles, S.E.; Kinghorn, J.R.; Tudzynski, B. 3-Hydroxy-3-methylglutaryl-CoA reductase gene of *Gibberella fujikuroi*: Isolation and characterization. *Curr. Genet.* **1997**, *31*, 38–47. [CrossRef] [PubMed]
25. Homann, V.; Mende, K.; Arntz, C.; Ilardi, V.; Macino, G.; Morelli, G.; Bose, G.; Tudzynski, B. The isoprenoid pathway: Cloning and characterization of fungal FPPS genes. *Curr. Genet.* **1996**, *30*, 232–239. [CrossRef] [PubMed]
26. Mende, K.; Homann, V.; Tudzynski, B. The geranylgeranyl diphosphate synthase gene of *Gibberella fujikuroi*: Isolation and expression. *Mol. Gen. Genet.* **1997**, *255*, 96–105. [CrossRef] [PubMed]
27. Tudzynski, B. Gibberellin biosynthesis in fungi: Genes, enzymes, evolution, and impact on biotechnology. *Appl. Microbiol. Biotechnol.* **2005**, *66*, 597–611. [CrossRef] [PubMed]
28. Fernández-Martín, R.; Cerdá-Olmedo, E.; Avalos, J. Homologous recombination and allele replacement in transformants of *Fusarium fujikuroi. Mol. Gen. Genet.* **2000**, *263*, 838–845. [CrossRef] [PubMed]
29. Schmidhauser, T.J.; Lauter, F.R.; Russo, V.E.; Yanofsky, C. Cloning, sequence, and photoregulation of *al-1*, a carotenoid biosynthetic gene of *Neurospora crassa. Mol. Cell. Biol.* **1990**, *10*, 5064–5070. [CrossRef] [PubMed]
30. Prado-Cabrero, A.; Schaub, P.; Díaz-Sánchez, V.; Estrada, A.F.; Al-Babili, S.; Avalos, J. Deviation of the neurosporaxanthin pathway towards β-carotene biosynthesis in *Fusarium fujikuroi* by a point mutation in the phytoene desaturase gene. *FEBS J.* **2009**, *276*, 4582–4597. [CrossRef] [PubMed]
31. Hausmann, A.; Sandmann, G. A single five-step desaturase is involved in the carotenoid biosynthesis pathway to β-carotene and torulene in *Neurospora crassa. Fungal Genet. Biol.* **2000**, *30*, 147–153. [CrossRef] [PubMed]
32. Schmidhauser, T.J.; Lauter, F.R.; Schumacher, M.; Zhou, W.; Russo, V.E.; Yanofsky, C. Characterization of *al-2*, the phytoene synthase gene of *Neurospora crassa*. Cloning, sequence analysis, and photoregulation. *J. Biol. Chem.* **1994**, *269*, 12060–12066. [PubMed]
33. Linnemannstöns, P.; Prado, M.M.; Fernández-Martín, R.; Tudzynski, B.; Avalos, J. A carotenoid biosynthesis gene cluster in *Fusarium fujikuroi*: the genes *carB* and *carRA. Mol. Genet. Genomics* **2002**, *267*, 593–602. [CrossRef] [PubMed]

34. Verdoes, J.C.; Krubasik, P.; Sandmann, G.; van Ooyen, A.J.J. Isolation and functional characterisation of a novel type of carotenoid biosynthetic gene from *Xanthophyllomyces dendrorhous*. *Mol. Gen. Genet.* **1999**, *262*, 453–461. [CrossRef] [PubMed]

35. Arrach, N.; Schmidhauser, T.J.; Avalos, J. Mutants of the carotene cyclase domain of *al-2* from *Neurospora crassa*. *Mol. Genet. Genom.* **2002**, *266*, 914–921.

36. Díaz-Sánchez, V.; Estrada, A.F.; Trautmann, D.; Limón, M.C.; Al-Babili, S.; Avalos, J. Analysis of *al-2* mutations in *Neurospora*. *PLoS ONE* **2011**, *6*, e21948. [CrossRef] [PubMed]

37. Sui, X.; Kiser, P.D.; von Lintig, J.; Palczewski, K. Structural basis of carotenoid cleavage: From bacteria to mammals. *Arch. Biochem. Biophys.* **2013**, *539*, 203–213. [CrossRef] [PubMed]

38. Thewes, S.; Prado-Cabrero, A.; Prado, M.M.; Tudzynski, B.; Avalos, J. Characterization of a gene in the car cluster of *Fusarium fujikuroi* that codes for a protein of the carotenoid oxygenase family. *Mol. Genet. Genom.* **2005**, *274*, 217–228. [CrossRef] [PubMed]

39. Prado-Cabrero, A.; Scherzinger, D.; Avalos, J.; Al-Babili, S. Retinal biosynthesis in fungi: Characterization of the carotenoid oxygenase CarX from *Fusarium fujikuroi*. *Eukaryot. Cell* **2007**, *6*, 650–657. [CrossRef] [PubMed]

40. Estrada, A.F.; Avalos, J. Regulation and targeted mutation of *opsA*, coding for the NOP-1 opsin orthologue in *Fusarium fujikuroi*. *J. Mol. Biol.* **2009**, *387*, 59–73. [CrossRef] [PubMed]

41. Díaz-Sánchez, V.; Limón, M.C.; Schaub, P.; Al-Babili, S.; Avalos, J. A RALDH-like enzyme involved in *Fusarium verticillioides* development. *Fungal Genet. Biol.* **2016**, *86*, 20–32. [CrossRef] [PubMed]

42. Prado-Cabrero, A.; Estrada, A.F.; Al-Babili, S.; Avalos, J. Identification and biochemical characterization of a novel carotenoid oxygenase: Elucidation of the cleavage step in the *Fusarium* carotenoid pathway. *Mol. Microbiol.* **2007**, *64*, 448–460. [CrossRef] [PubMed]

43. Saelices, L.; Youssar, L.; Holdermann, I.; Al-Babili, S.; Avalos, J. Identification of the gene responsible for torulene cleavage in the *Neurospora* carotenoid pathway. *Mol. Genet. Genom.* **2007**, *278*, 527–537. [CrossRef] [PubMed]

44. Jin, J.M.; Lee, J.; Lee, Y.W. Characterization of carotenoid biosynthetic genes in the ascomycete *Gibberella zeae*. *FEMS Microbiol. Lett.* **2010**, *302*, 197–202. [CrossRef] [PubMed]

45. Goldie, A.H.; Subden, R.E. The neutral carotenoids of wild-type and mutant strains of *Neurospora crassa*. *Biochem. Genet.* **1973**, *10*, 275–284. [CrossRef] [PubMed]

46. Sandmann, G. Photoregulation of carotenoid biosynthesis in mutants of *Neurospora crassa*: Activities of enzymes involved in the synthesis and conversion of phytoene. *Z. Naturforsch.* **1993**, *48c*, 570–574.

47. Estrada, A.F.; Youssar, L.; Scherzinger, D.; Al-Babili, S.; Avalos, J. The *ylo-1* gene encodes an aldehyde dehydrogenase responsible for the last reaction in the *Neurospora* carotenoid pathway. *Mol. Microbiol.* **2008**, *69*, 1207–1220. [CrossRef] [PubMed]

48. Díaz-Sánchez, V.; Estrada, A.F.; Trautmann, D.; Al-Babili, S.; Avalos, J. The gene *carD* encodes the aldehyde dehydrogenase responsible for neurosporaxanthin biosynthesis in *Fusarium fujikuroi*. *FEBS J.* **2011**, *278*, 3164–3176. [CrossRef] [PubMed]

49. Rau, W. Über den Einfluss der Temperatur auf die lichtabhängige Carotinoidbildung von *Fusarium aquaeductuum*. *Planta* **1962**, *59*, 123–137. [CrossRef]

50. Rau, W. Untersuchungen über die lichtabhängige Carotinoidsynthese. II. Ersatz der Lichtinduktion durch Mercuribenzoat. *Planta* **1967**, *74*, 263–277. [CrossRef] [PubMed]

51. Rau, W. Untersuchungen über die lichtabhängige Carotinoidsynthese. VII. Reversible Unterbrechung der Reaktionskette durch Cycloheximid und anaerobe Bedingungen. *Planta* **1971**, *101*, 251–264. [CrossRef] [PubMed]

52. Rau, W.; Feuser, B.; Rau-Hund, A. Substitution of p-chloro- or p-hydroxymercuribenzoate for light in carotenoid synthesis by *Fusarium aquaeductuum*. *Biochim. Biophys. Acta* **1967**, *136*, 589–590. [CrossRef]

53. Theimer, R.R.; Rau, W. Untersuchungen über die lichtabhängige Carotinoidsynthese V. Aufhebung der Lichtinduktion dutch Reduktionsmittel und Ersatz des Lichts durch Wasserstoffperoxid. *Planta* **1970**, *92*, 129–137. [CrossRef] [PubMed]

54. Theimer, R.R.; Rau, W. Untersuchungen über die lichtabhängige Carotinoidsynthese VIII. Die unterschiedlichen Wirkungsmechanismen von Licht und Mercuribenzoat. *Planta* **1972**, *106*, 331–343. [CrossRef] [PubMed]

55. Avalos, J.; Cerdá-Olmedo, E. Chemical modification of carotenogenesis in *Gibberella fujikuroi*. *Phytochemistry* **1986**, *25*, 1837–1841. [CrossRef]

56. Rau, W. Untersuchungen über die lichtabhängige Carotinoidsynthese. I. Das Wirkungsspektrum von *Fusarium aquaeductuum*. *Planta* **1967**, *72*, 14–28. [CrossRef] [PubMed]

57. Schrott, E.L.; Huber-Willer, A.; Rau, W. Is phytochrome involved in the light-mediated carotenogenesis in *Fusarium aquaeductuum* and *Neurospora crassa*? *Photochem. Photobiol.* **1982**, *35*, 213–216. [CrossRef]

58. Lang-Feulner, J.; Rau, W. Redox dyes as artificial photoreceptors in light-dependent carotenoid synthesis. *Photochem. Photobiol.* **1975**, *21*, 179–183. [CrossRef] [PubMed]

59. Prado, M.M.; Prado-Cabrero, A.; Fernández-Martín, R.; Avalos, J. A gene of the opsin family in the carotenoid gene cluster of *Fusarium fujikuroi*. *Curr. Genet.* **2004**, *46*, 47–58. [CrossRef] [PubMed]

60. De Fabo, E.C.; Harding, R.W.; Shropshire, W., Jr. Action spectrum between 260 and 800 nanometers for the photoinduction of carotenoid biosynthesis in *Neurospora crassa*. *Plant Physiol.* **1976**, *57*, 440–445. [CrossRef] [PubMed]

61. Fischer, R.; Aguirre, J.; Herrera-Estrella, A.; Corrochano, L.M. The complexity of fungal vision. *Microbiol. Spectr.* **2016**, *4*, 1–22.

62. Estrada, A.F.; Avalos, J. The White Collar protein WcoA of *Fusarium fujikuroi* is not essential for photocarotenogenesis, but is involved in the regulation of secondary metabolism and conidiation. *Fungal Genet. Biol.* **2008**, *45*, 705–718. [CrossRef] [PubMed]

63. Ruiz-Roldán, M.C.; Garre, V.; Guarro, J.; Mariné, M.; Roncero, M.I. Role of the white collar 1 photoreceptor in carotenogenesis, UV resistance, hydrophobicity, and virulence of *Fusarium oxysporum*. *Eukaryot. Cell* **2008**, *7*, 1227–1230. [CrossRef] [PubMed]

64. Kim, H.; Son, H.; Lee, Y.-W. Effects of light on secondary metabolism and fungal development of *Fusarium graminearum*. *J. Appl. Microbiol.* **2014**, *116*, 380–389. [CrossRef] [PubMed]

65. Castrillo, M.; Avalos, J. The flavoproteins CryD and VvdA cooperate with the white collar protein WcoA in the control of photocarotenogenesis in *Fusarium fujikuroi*. *PLoS ONE* **2015**, *10*, e0119785. [CrossRef] [PubMed]

66. Castrillo, M.; Bernhardt, A.; Avalos, J.; Batschauer, A.; Pokorny, R. Biochemical characterization of the DASH-type cryptochrome CryD from *Fusarium fujikuroi*. *Photochem. Photobiol.* **2015**, *91*, 1356–1367. [CrossRef] [PubMed]

67. Castrillo, M.; García-Martínez, J.; Avalos, J. Light-dependent functions of the *Fusarium fujikuroi* CryD DASH cryptochrome in development and secondary metabolism. *Appl. Environ. Microbiol.* **2013**, *79*, 2777–2788. [CrossRef] [PubMed]

68. Rodríguez-Ortiz, R.; Michielse, C.; Rep, M.; Limón, M.C.; Avalos, J. Genetic basis of carotenoid overproduction in *Fusarium oxysporum*. *Fungal Genet. Biol.* **2012**, *49*, 684–696. [CrossRef] [PubMed]

69. Ádám, A.L.; García-Martínez, J.; Szücs, E.P.; Avalos, J.; Hornok, L. The MAT1-2-1 mating-type gene upregulates photo-inducible carotenoid biosynthesis in *Fusarium verticillioides*. *FEMS Microbiol. Lett.* **2011**, *318*, 76–83. [CrossRef] [PubMed]

70. Avalos, J.; Corrochano, L.M. Carotenoid biosynthesis in *Neurospora*. In *Neurospora: Genomics and Molecular Biology*; Kasbekar, D.P., McCluskey, K., Eds.; Caister Academic Press: Norfolk, UK, 2013; pp. 227–241.

71. Ruger-Herreros, M. Participación de la Proteína CarS en la Regulación de la Carotenogénesis y el Estrés en *Fusarium fujikuroi*. Ph.D. Thesis, Universidad de Sevilla, Seville, Spain, 2016.

72. Castrillo, M.; Avalos, J. Light-mediated participation of the VIVID-like protein of *Fusarium fujikuroi* VvdA in pigmentation and development. *Fungal Genet. Biol.* **2014**, *71*, 9–20. [CrossRef] [PubMed]

73. Avalos, J.; Casadesús, J.; Cerdá-Olmedo, E. *Gibberella fujikuroi* mutants obtained with UV radiation and N-methyl-N'-nitro-N-nitrosoguanidine. *Appl. Environ. Microbiol.* **1985**, *49*, 187–191. [PubMed]

74. Avalos, J.; Mackenzie, A.; Nelki, D.S.; Bramley, P.M. Terpenoid biosynthesis in cell-extracts of wild type and mutant strains of *Gibberella fujikuroi*. *Biochim. Biophys. Acta* **1988**, *966*, 257–265. [CrossRef]

75. Delgado-Pelayo, R.; Hornero-Méndez, D. Identification and quantitative analysis of carotenoids and their esters from sarsaparrilla (*Smilax aspera* L.) berries. *J. Agric. Food Chem.* **2012**, *60*, 8225–8232. [CrossRef] [PubMed]

76. Rodríguez-Ortiz, R.; Limón, M.C.; Avalos, J. Functional analysis of the *carS* gene of *Fusarium fujikuroi*. *Mol. Genet. Genom.* **2013**, *288*, 157–173. [CrossRef] [PubMed]

77. Navarro, E.; Lorca-Pascual, J.M.; Quiles-Rosillo, M.D.; Nicolás, F.E.; Garre, V.; Torres-Martínez, S.; Ruiz-Vázquez, R.M. A negative regulator of light-inducible carotenogenesis in *Mucor circinelloides*. *Mol. Genet. Genom.* **2001**, *266*, 463–470.

78. Lorca-Pascual, J.M.; Murcia-Flores, L.; Garre, V.; Torres-Martínez, S.; Ruiz-Vázquez, R.M. The RING-finger domain of the fungal repressor *crgA* is essential for accurate light regulation of carotenogenesis. *Mol. Microbiol.* **2004**, *52*, 1463–1474. [CrossRef] [PubMed]

79. Navarro, E.; Ruiz-Pérez, V.L.; Torres-Martínez, S. Overexpression of the *crgA* gene abolishes light requirement for carotenoid biosynthesis in *Mucor circinelloides*. *Eur. J. Biochem.* **2000**, *267*, 800–807. [CrossRef] [PubMed]

80. Silva, F.; Navarro, E.; Peñaranda, A.; Murcia-Flores, L.; Torres-Martínez, S.; Garre, V. A RING-finger protein regulates carotenogenesis via proteolysis-independent ubiquitylation of a White Collar-1-like activator. *Mol. Microbiol.* **2008**, *70*, 1026–1036. [CrossRef] [PubMed]

81. Quiles-Rosillo, M.D.; Torres-Martínez, S.; Garre, V. *Ciga*, a light-inducible gene involved in vegetative growth in *Mucor circinelloides* is regulated by the carotenogenic repressor *crgA*. *Fungal Genet. Biol.* **2003**, *38*, 122–132. [CrossRef]

82. Murcia-Flores, L.; Lorca-Pascual, J.M.; Garre, V.; Torres-Martínez, S.; Ruiz-Vázquez, R.M. Non-AUG translation initiation of a fungal RING finger repressor involved in photocarotenogenesis. *J. Biol. Chem.* **2007**, *282*, 15394–15403. [CrossRef] [PubMed]

83. Candau, R.; Avalos, J.; Cerdá-Olmedo, E. Gibberellins and carotenoids in the wild type and mutants of *Gibberella fujikuroi*. *Appl. Environ. Microbiol.* **1991**, *57*, 3378–3382. [PubMed]

84. Rodríguez-Ortiz, R.; Limón, M.C.; Avalos, J. Regulation of carotenogenesis and secondary metabolism by nitrogen in wild-type *Fusarium fujikuroi* and carotenoid-overproducing mutants. *Appl. Environ. Microbiol.* **2009**, *75*, 405–413. [CrossRef] [PubMed]

85. Studt, L.; Tudzynski, B. Gibberellins and the red pigments bikaverin and fusarubin. In *Biosynthesis and Molecular Genetics of Fungal Secondary Metabolites*; Martín, J.-F., García-Estrada, C., Zeilinger, S., Eds.; Springer: New York, NY, USA, 2014; pp. 209–238, ISBN 978-1-4939-1190-5.

86. Niehaus, E.-M.; Díaz-Sánchez, V.; von Bargen, K.W.; Kleigrewe, K.; Humpf, H.-U.; Limón, M.C.; Tudzynski, B. Fusarins and Fusaric Acid in Fusaria. In *Biosynthesis and Molecular Genetics of Fungal Secondary Metabolites*; Martín, J.-F., García-Estrada, C., Zeilinger, S., Eds.; Springer: New York, NY, USA, 2014; pp. 239–262, ISBN 978-1-4939-1190-5.

87. Garbayo, I.; Vílchez, C.; Nava-Saucedo, J.E.; Barbotin, J.N. Nitrogen, carbon and light-mediated regulation studies of carotenoid biosynthesis in immobilized mycelia of *Gibberella fujikuroi*. *Enzyme Microb. Technol.* **2003**, *33*, 629–634. [CrossRef]

88. Sokolovsky, V.Y.; Lauter, F.R.; Müller-Röber, B.; Ricci, M.; Schmidhauser, T.J.; Russo, V.E.A. Nitrogen regulation of blue light-inducible genes in *Neurospora crassa*. *J. Gen. Microbiol.* **1992**, *138*, 2045–2049. [CrossRef]

89. Connolly, L.R.; Smith, K.M.; Freitag, M. The *Fusarium graminearum* histone H3 K27 methyltransferase KMT6 regulates development and expression of secondary metabolite gene clusters. *PLoS Genet.* **2013**, *9*, e1003916. [CrossRef] [PubMed]

90. Wiemann, P.; Sieber, C.M.; von Bargen, K.W.; Studt, L.; Niehaus, E.M.; Espino, J.J.; Huss, K.; Michielse, C.B.; Albermann, S.; Wagner, D.; et al. Deciphering the cryptic genome: Genome-wide analyses of the rice pathogen *Fusarium fujikuroi* reveal complex regulation of secondary metabolism and novel metabolites. *PLoS Pathog.* **2013**, *9*, e1003475. [CrossRef] [PubMed]

91. García-Martínez, J.; Ádám, A.L.; Avalos, J. Adenylyl cyclase plays a regulatory role in development, stress resistance and secondary metabolism in *Fusarium fujikuroi*. *PLoS ONE* **2012**, *7*, e28849. [CrossRef] [PubMed]

92. Studt, L.; Humpf, H.-U.; Tudzynski, B. Signaling governed by G proteins and cAMP is crucial for growth, secondary metabolism and sexual development in *Fusarium fujikuroi*. *PLoS ONE* **2013**, *8*, e58185. [CrossRef] [PubMed]

93. Rodríguez-Ortiz, R. Análisis Genético y Molecular del Fenotipo CarS en *Fusarium*. Ph.D. Thesis, Universidad de Sevilla, Seville, Spain, 2012.

94. Domenech, C.E.; Giordano, W.; Avalos, J.; Cerdá-Olmedo, E. Separate compartments for the production of sterols, carotenoids and gibberellins in *Gibberella fujikuroi*. *Eur. J. Biochem.* **1996**, *239*, 720–725. [CrossRef] [PubMed]

Journal of
Fungi

MDPI

Review

Biosynthesis of Astaxanthin as a Main Carotenoid in the Heterobasidiomycetous Yeast *Xanthophyllomyces dendrorhous*

Jose L. Barredo [1], Carlos García-Estrada [2,3], Katarina Kosalkova [2] and Carlos Barreiro [2,4,*]

[1] CRYSTAL PHARMA S.A.U. Parque Tecnológico de León, C/Nicostrato Vela s/n, 24009 León, Spain;
    JoseLuis.Barredo@amriglobal.com
[2] INBIOTEC (Instituto de Biotecnología de León), Avda. Real, 1, 24006 León, Spain;
    cgare@unileon.es (C.G.-E.); kkos@unileon.es (K.K.)
[3] Área de Toxicología, Departamento de Ciencias Biomédicas, Universidad de León, Campus de Vegazana,
    24071 León, Spain
[4] Área de Microbiología, Departamento de Biología Molecular, Universidad de León, Campus de Ponferrada,
    Avda, Astorga, s/n, 24400 Ponferrada, Spain
*   Correspondence: c.barreiro@unileon.es; Tel.: +34-987-210-308; Fax: +34-987-210-388

Received: 13 June 2017; Accepted: 27 July 2017; Published: 30 July 2017

Abstract: Carotenoids are organic lipophilic yellow to orange and reddish pigments of terpenoid nature that are usually composed of eight isoprene units. This group of secondary metabolites includes carotenes and xanthophylls, which can be naturally obtained from photosynthetic organisms, some fungi, and bacteria. One of the microorganisms able to synthesise carotenoids is the heterobasidiomycetous yeast *Xanthophyllomyces dendrorhous*, which represents the teleomorphic state of *Phaffia rhodozyma*, and is mainly used for the production of the xanthophyll astaxanthin. Upgraded knowledge on the biosynthetic pathway of the main carotenoids synthesised by *X. dendrorhous*, the biotechnology-based improvement of astaxanthin production, as well as the current omics approaches available in this yeast are reviewed in depth.

Keywords: *Xanthophyllomyces dendrorhous; Phaffia rhodozyma;* astaxanthin; carotenoids; carotenes; xanthophylls

---

1. Introduction: *Xanthophyllomyces dendrorhous* and Carotenoids

The carotenoids group includes tetraterpenoid organic pigments, the majority comprising of eight isoprene units with a C40 skeleton. These lipophilic metabolites are insoluble in water and contain a long polyene central chain of conjugated double bonds that functions as a chromophore (400–500 nanometers being the electromagnetic spectrum where carotenoids absorb maximally) responsible for the characteristic yellow to orange and reddish colours of these compounds [1].

All naturally occurring carotenoids are produced by photosynthetic species (including plants and algae), and by some classes of fungi and non-photosynthetic bacteria [2–5]. In general, animals are unable to produce their own carotenoids and therefore, the only way to obtain these compounds is from their diet.

Carotenoids can be classified according to the oxygenation degree into carotenes and xanthophylls. Carotenes (e.g., β-carotene, α-carotene or lycopene) are strictly hydrocarbons (non-oxygenated molecules), whereas xanthophylls (e.g., lutein, zeaxanthin, canthaxanthin, or astaxanthin) are oxygenated molecules (oxycarotenoids) with a hydroxy, epoxy, and/or oxo group [6]. These compounds play different roles. Thus, in photosynthesizing species they are associated with the light harvesting complexes acting as accessory light-harvesting pigments, effectively extending the range of light absorbed by the photosynthetic apparatus. In those organisms, carotenoids also play a photoprotective role by quenching triplet state chlorophyll molecules and scavenging singlet oxygen and other toxic oxygen species formed within the chloroplast, and in the case of zeaxanthin, by dissipating harmful excess excitation energy under stress conditions [7,8]. Besides, carotenoids provide organisms of bright yellow, red, or orange and their main function in all non-photosynthetic organisms seems to be (photo) protection. They are known to be very efficient physical and chemical quenchers of singlet oxygen ($^1O_2$), as well as potent scavengers of other reactive oxygen species, playing an important role as antioxidants [9,10]. Carotenoids are also important precursors of retinol (vitamin A precursors) [11].

More than 700 types of carotenoids have been found from natural sources so far [5,12]. The carotenoids market in 2019/2020 is supposed to reach $1.5–1.8 billion with a compound annual growth rate of 3.9% [13,14] and due to the extensive commercial and industrial uses of carotenoids (mainly lutein and astaxanthin), the demand of these compounds is high around the world. Therefore, several biological platforms are used for the biotechnological production of natural carotenoids for their use in food and feed, cosmetics, and the chemical and pharmaceutical industries [15].

One of these platforms is the red/pink-pigmented heterobasidiomycetous yeast *Xanthophyllomyces dendrorhous* (the teleomorphic state of *Phaffia rhodozyma*), which was isolated in the late 1960s from tree-exudates in Japan and Alaska [16] and naturally produces and accumulates the xanthophyll astaxanthin [17]. This yeast can be considered as a cell-factory for the production of industrially valuable carotenoids [18], since the genome of *X. dendrorhous* CBS6938 has been sequenced [19] and biotechnology tools for the genetic manipulations of this microorganism are available [20–23]. In addition, this yeast does not require light for accumulation of astaxanthin, is able to metabolise many kinds of saccharides under both aerobic and anaerobic conditions, and reproduces at relatively high growth rates [24]. Besides, its approval as a colour stabiliser for fish feed supplementation by FDA (U.S. Food and Drug Administration) supports the natural production as an interesting methodology (https://www.accessdata.fda.gov/scripts/cdrh/cfdocs/cfcfr/CFRSearch.cfm?fr=73.355).

In addition to astaxanthin, *X. dendrorhous* is able to produce several carotenoids, including β-carotene, canthaxanthin, zeaxanthin, and astaxanthin via mevalonate pathway (Figure 1). β-carotene, a red-orange carotenoid, possesses β-rings at both ends and serves as an intermediary molecule for the biosynthesis of astaxanthin [25,26]. This compound is widely used in the food, feed, cosmetic, and pharmaceutical industries due to its potent colouring traits, antioxidant properties, and provitamin A activity, since it appears to be the most important vitamin A precursor for vertebrate animal species [27]. Besides, β-carotene adds colour to beverages, dairy products, confectionery, and many other commodities, and together with astaxanthin, they are the most important carotenoids from a commercial point of view. The orange-red pigment canthaxanthin (β,β-carotene-4,4′-dione) is a xanthophyll with antioxidant properties widely used in aquaculture and poultry farming, providing the characteristic colour to fish flesh, chicken skin, and egg yolk [28,29]. Zeaxanthin (β,β-carotene-3,3′-diol) is a yellow xanthophyll alcohol important for vision, since together with lutein and meso-zeaxanthin, they are present in high concentrations within the oval-shaped macular pigmented area near the centre of the retina of the eye [30,31], thus playing an important protective role against several eye diseases [32]. Astaxanthin (3,3′-dihydroxy-β,β-carotene-4,4′-dione) is a red-orange pigment with a market value ranging from $2500–7000/kg whose global market was valued at US$447 in 2014 and expected to reach a value of US$1.1 billion by 2020 [14,33]. After β-carotene, astaxanthin is the second most important carotenoid, representing about 29% of total carotenoid sales [34]. This pigment has strong antioxidant properties and is used as a feed additive in salmon and trout aquacultures as well as in chicken and quail farming and egg production [6,35,36]. Other properties have been described for astaxanthin, such as beneficial effects in cardiovascular, immune, inflammatory, diabetes, carcinogenic, and neurodegenerative diseases, and as an antiaging and sun proofing agent [15,37–41]. The commercial origin of astaxanthin is from either chemical synthesis or natural resources such as fermentative production (microalgae, yeast) and crustacean byproducts [36,41]. Therefore, more than 95% of the global market refers to synthetically obtained astaxanthin, which presents lower production costs (around $1000/kg) than the biological alternative. Nowadays, the major manufacturers are DSM (The Netherlands), BASF (France), and NHU (China). However, its petrochemical origin limits the final use, which is boosting the natural sources of this carotenoid [14,41]. Thus, astaxanthin from natural sources, including bacteria such as *Paracoccus carotinifaciens*, yeasts like *X. dendrorhous*, or algae like *Haematococcus pluvialis*, is a realistic alternative to synthetic astaxanthin.

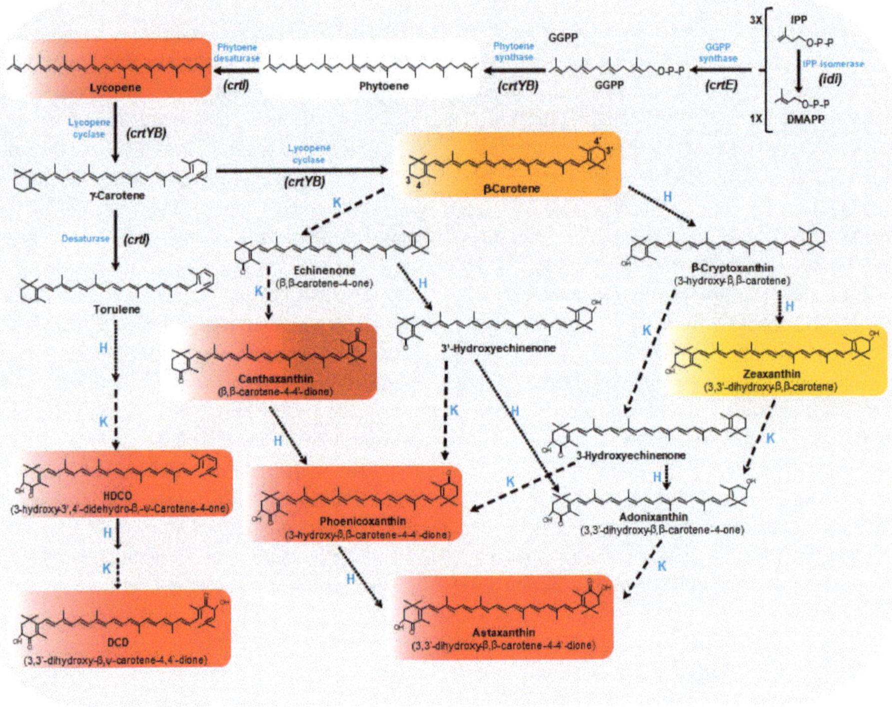

Figure 1. Astaxanthin biosynthetic pathway in *X. dendrorhous*. Initially, a molecule of dimethylallyl pyrophosphate (DMAPP) and three molecules of isopentenyl pyrophosphate (IPP) are combined to obtain geranylgeranyl pyrophosphate (GGPP) by means of the GGPP synthase. Secondly, two molecules of GGPP are coupled by the phytoene synthase (*crtYB* gene) to reach phytoene. The phytoene desaturase(*crtI* gene) introduces four double bonds in the molecule of phytoene to to obtain lycopene. Then, the lycopene cyclase (*crtYB* gene) converts one of the ψ acyclic ends of lycopene as β-ring to form γ-carotene, and subsequently the other to form β-carotene. Xanthophylls bioconversion in *X. dendrorhous* from β-carotene and γ-carotene includes the addition of two 4-keto groups in the molecule of β-carotene by the ketolase (K, discontinuous line) activity, and the inclusion of two 3-hydroxy groups by the hydroxylase (H, continuous dotted line) activity. Both K and H activities are presented in a single enzyme (astaxanthin synthetase; CrtS) encoded by a single gene (*crtS*). The cytochrome P450 reductase encoded by the *crtR* gene is a CrtS helper protein, providing with electrons for substrate oxygenation. The existence of a monocyclic pathway to DCD was also proposed [42]. Main carotenoids detected in *X. dendrorhous* broths are shown inside a rectangle in concordance with their natural colours. This figure has been based on Rodríguez-Sáiz and co-workers [43].

The aim of this review is to provide an up-to-date version of the the carotenoids biosynthetic pathway of *X. dendrorhous*, together with the main biotechnology approaches and omics tools that have been applied to this yeast for the improvement of astaxanthin production.

2. The Biosynthetic Pathway of Astaxanthin in *X. dendrorhous*

The astaxanthin biosynthetic pathway is encoded by six genes (*crtI*, *crtL*, *crtR*, *crtO*, *crtW*, and *crtZ*), which present an evolutionary pattern (eukaryotic and prokaryotic) characterised by lateral gene transfer and gene duplication events. These genes are more conserved in plants and algae than in any other bacterial phyla, where the structural genes evolve slower than the regulatory genes [44]. This biosynthetic pathway (Figure 1) has been widely studied in *X. dendrorhous* [45–48]. Thus, the biosynthesis begins with a C5 isoprene unit to which prenyl transferases sequentially add three other isoprenic units [49], resulting in the formation of C20 geranylgeranyl-pyrophosphate (GGPP). The active forms of the isoprene unit are isopentenyl-pyrophosphate (IPP) and its allylic isomer dimethylallyl-pyrophosphate (DMAPP). In most eukaryotes, IPP derives from the mevalonate pathway [50], while in prokaryotes and in plant plastids, it is synthesised via the 2-C-methyl-D-erythritol-4-phosphate (MEP) pathway, which is also known as the non-mevalonate pathway [51].

The *idi* gene encodes IPP isomerase, which catalyses the isomerisation of IPP to DMAPP [52], and then both molecules are joined together, generating C10-geranyl pyrophosphate (GPP), the precursor of monoterpenes [53]. The addition of a second molecule of IPP to GPP by prenyl transferases gives the C15 precursor of sesquiterpenes, farnesyl pyrophosphate (FPP), which is converted into GGPP [54] by the further addition of IPP by the GGPP synthase (encoded by the *crtE* gene). Next, phytoene synthase (encoded by the *crtYB* gene) links two molecules of GGPP in a tail-to-tail manner, yielding phytoene [50,55]. This is the first carotenoid synthesised in the pathway, which is colourless as it has a symmetrical carotenoid skeleton with only three conjugated double bonds. The structural diversity of carotenoids is generated by further modifications, including desaturations, cyclisations, isomerisations, and oxygenations [56]. The phytoene synthase of fungi is a bifunctional enzyme that has both phytoene synthase and lycopene cyclase activities, which gives rise to β-carotene. In this particular enzyme, encoded by the *crtYB* gene, the phytoene synthase and lycopene cyclase activities are located in the C-terminal and N-terminal functional domains, respectively.

Next, phytoene is desaturated by phytoene desaturase (encoded by the *crtI* gene) [3,57] by the incorporation of two, three, four, or five double bonds, thus producing the coloured carotenoids: (i) ζ-carotene (yellow, synthesised by some plants, cyanobacteria, and algae); (ii) neurosporene (yellow, accumulates in *Rhodobacter capsulatus* and *R. sphaeroides*); (iii) lycopene (red, found in most eubacteria and fungi) or (iv) 3,4-didehydrolycopene (found in the mold *Neurospora crassa*) [50], respectively.

Although there are acyclic carotenoids, the cyclisation of lycopene is a frequent step in the biosynthesis of carotenoids, forming three types of ionone rings: β-, ε-, and γ(gamma)-rings (Britton 1998). The β-ring is the most common, the ε-ring is found in plants and in some algae, and the γ-ring is the rarest one. The lycopene cyclase (also encoded by the *crtYB* gene) [58] sequentially converts the ψ acyclic ends of lycopene to β rings to form γ-carotene and β-carotene [55].

The synthesis of xanthophylls involves the oxidation of post-phytoene carotenoid molecules, mainly from α- and β-carotenes, resulting in oxygenated products with hydroxyl-, epoxy-, and oxo-functional groups. The formation of astaxanthin from β-carotene involves the introduction of a hydroxyl and a keto group at C3 and C4, respectively, for each of the β-ionone rings. Two enzymatic activities convert β-carotene into astaxanthin through several biosynthetic intermediates: a ketolase, which incorporates two 4-keto groups in the molecule of β-carotene, and a hydroxylase, which introduces two 3-hydroxy groups. In *X. dendrorhous*, both activities are included in a single enzyme (astaxanthin synthetase; CrtS), belonging to the cytochrome P450 family and encoded by the *crtS* gene, which sequentially catalyses 4-ketolation of β-carotene followed by 3-hydroxylation [25,26]. In contrast, two independent genes have been described in other astaxanthin producing microorganisms.

A cytochrome P450 reductase, encoded by the *crtR* gene, has been shown to have an auxiliary role to CrtS in *X. dendrorhous*, providing the necessary electrons for substrate oxygenation [59]. A mutant of *X. dendrorhous* lacking the *crtR* gene accumulates β-carotene and is unable to synthesise astaxanthin, demonstrating that CrtR is essential for the synthesis of astaxanthin [59]. In *Saccharomyces cerevisiae* strains expressing *X. dendrorhous* carotenogenic genes, astaxanthin production was only achieved when CrtS was co-expressed with CrtR [60].

The existence of a monocyclic pathway diverging from the dicyclic pathway at neurosporene and proceeding through β-zeacarotene, 3,4-didehydrolycopene, torulene, 3-hydroxy-3′,4′-didehydro-β, ψ-carotene-4-one (HDCO) to the end product 3,3′-dihydroxy-β, ψ-carotene-4,4′-dione (DCD) (Figure 1) was also proposed [42].

3. Biotechnology-Based Improvement of Astaxanthin Production in *X. dendrorhous*

Chemical synthesis is relatively complex even for simple carotenoids such as β-carotene. Thus, several companies and academic laboratories have investigated biological sources of astaxanthin. Among these sources, *P. rhodozyma*/*X. dendrorhous* and the microalga *H. pluvialis* stand out because of their ability to synthesise astaxanthin. However, biological production of astaxanthin in natural isolates presents a well-known drawback; the very low specific production. Thus, the specific astaxanthin production of wild type strains of *X. dendrorhous* is 200–400 μg/g of yeast dry weight [61,62]. Besides, the thickness of yeast cell walls and capsule, which apparently hinders astaxanthin uptake, seems to be also relevant for production and downstream processing. In order to obtain a competitive natural source of this xanthophyll it has been necessary to increase this production by a factor of 10–50. Classical random mutagenesis methods by using physical or chemicals agents have been successfully applied [34,43,63–65]. Some of these mutants obtained by random mutagenesis and screening with relevant dry cell weight productivity rates [*P. rhodozyma* E5042 (2.5 mg/g); *X. dendrorhous* VKPM Y2476 (4.1 mg/g); *P. rhodozyma* JMU-MVP114 (6.01 mg/g)] [66–68]. Nevertheless, one of the drawbacks of this approach is the introduction of secondary mutations that may affect the physiology, viability, or metabolic capacity of the cell. Besides, genetic instability of the mutants, reduction in the production of biomass, and accumulation of undesired intermediaries have frequently been detected [64,69]. The selection of astaxanthin super-producer mutants in solid medium conditioned the detection of strains due to the low levels of aeration, which inhibit the production of oxygenated carotenoids. As a result, different selection procedures have been developed. On the one hand, positive selection by means of inhibitors of carotenogenesis (β-ionone, diphenylamine) has been developed [70]. On the other hand, negative selection, such as the increased sensitivity of carotenoid-producing strains to antimycin A (a respiratory inhibitor), has been another option [71,72].

The yield improvement of microbial products has been traditionally faced up by means of: (i) treatment with mutagenic agents (e.g., nitrogen mustards, ultraviolet irradiation, X-ray); (ii) increasing the biosynthetic genes; or (iii) redirecting the metabolic precursors [73]. As a consequence, a promising strategy is metabolic engineering (see Table 1), which includes: (i) the overexpression of carotenoid biosynthetic genes (e.g., *crtYB* gene); (ii) the metabolic flow increase towards the synthesis of specific pathway precursor molecules (e.g., geranylgeranyl-pyrophosphate synthase encoding gene (*crtE*)), or (iii) the supplementation with carotenogenesis precursors (e.g., mevalonate) [20,47,62,74,75]. However, this metabolic redirection, which can be included in the new trends of synthetic biology, has needed several methodological updates prior to validating the final effects along the astaxanthin production improvement history. As an example, transformation methods for *Xanthophyllomyces* have been developed and optimised using (i) constitutive promoters from the yeast itself [e.g., actin, glyceraldehyde-3-phosphate dehydrogenase (*gpd*) or NADP-dependent glutamate dehydrogenase (*gdhA*) genes]; (ii) stable integration in the genome by means of the ribosomal DNA of *X. dendrorhous* introduced into the transformation vectors; and (iii) optimised resistance genes (e.g., kanamycin resistance gene of transposon Tn5 or hygromycin resistance gene (*hph*)) [21,76]. Under optimum conditions, the transformation efficiency obtained was 1×10^3 transformants per microgram of plasmid DNA [76].

Table 1. Examples summary of *X. dendrorhous* (*P. rhodozyma*) genetic engineering.

Targets	Approach	Result	Ref.
crtYB gene	Deactivation	No carotenoids	[48]
crtYB gene	Overexpression	Accumulation of β-carotene and echinenone	[48]
crtI gene	Overexpression	Increase torulene and HDCO and decrease echinenone, L-carotene and astaxanthin	[48]
crtR gene	Description of its role	Required together with the *crtS* gene for the conversion of β-carotene to astaxanthin.	[59]
double *cyp61* genes	Deletion	Enhanced astaxanthin production by 1,4-fold compared with the parental strain	[77]
crtE gene under Padh4r	Evaluation of promotors	Increase in intracellular astaxanthin by 1.7-fold compared with parental	[78]
acaT, *hmgS* and *hmgR* genes	Triple overexpression	Enhanced volumetric astaxanthin production by 1.4-fold compared with that of the control strain	[23]
acaT/hmgS/hmgR/crtE/crtS genes	Combined overexpression	Enhanced volumetric astaxanthin production by 2.1-fold higher compared with the control strain	[23]
Combination of conventional mutagenesis and *crtYB* gene expression	Combined overexpression	22 times higher astaxanthin specific production than for the wild type	[75]

The carotenoid pathway of *X. dendrorhous* is well established (Figure 1). This fact has enabled the cloning process of all genes involved in phytoene synthesis, phytoene desaturation, and cyclisation and formation of 3-hydroxy and 4-keto. Thus, knowledge of the molecular biology of *X. dendrorhous* allows researchers to direct the genetic modifications and increase the metabolic precursors flow to the carotenoid biosynthetic pathway [48,59,79,80]. In order to divert metabolite flow from the sterol pathway towards carotenoid biosynthesis, *X. dendrorhous* was transformed with the *crtE* cDNA (geranylgeranyl pyrophosphate synthase). Transformants were obtained with higher carotenoid levels including astaxanthin due to increased levels of geranylgeranyl pyrophosphate synthase [74].

The genetic improvement of *Xanthophyllomyces* strains by metabolic engineering has been achieved by upregulating phytoene synthase and lycopene cyclase leading to an increase in β-carotene and echinenone titres. These results suggested that the oxygenation reactions might be rate limiting [48]. The combination of chemical mutagenesis and genetic engineering by targeting limiting reactions (including overexpression of *crtYB* and *crtS* genes) led to the generation of an overproducing strain of astaxanthin (9.7 mg per gram of dry weight) [75]. These authors described the construction of new transformation plasmids for the stepwise expression of the bottleneck genes in the carotenoid biosynthetic pathway. As a result, titres were comparable to those provided by *H. pluvialis*, the leading commercial producer of natural astaxanthin [22].

Enhancement of *crtS* gene transcriptional levels led to an increase in the transcription levels of related genes (*crtE*, *crtYB*, *crtI*) in the astaxanthin biosynthetic pathway. A scheme of carotenoid biosynthesis in *X. dendrorhous* involving alternative bicyclic and monocyclic pathways was proposed by Chi and co-workers in 2015 [81].

The limiting step in the carotenoid pathway is phytoene synthase. In order to increase the *crtYB* gene copy number, integration plasmids were constructed. Thus, the transformants with higher copy number accumulated carotenoid intermediates missed in the parental strain. Some of them are substrates and intermediates of astaxanthin synthase, which can be transformed to astaxanthin by the addition of the astaxanthin synthase gene [82].

Recently, a system to completely delete target diploid genes in *X. dendrorhous* was developed. Diploid *CYP61* genes involved in the synthesis of ergosterol that inhibits the pathway for mevalonate (substrate for isoprenoid biosynthesis), were deleted. Ergosterol biosynthesis was decreased, whereas astaxanthin production was approximately 1.4-fold higher than the parental strain [77].

Many reports have been published about astaxanthin fermentation in *X. dendrorhous*, but there are few reports about large-scale production. In order to produce astaxanthin by biotechnological processes, the scale-up step from lab scale to industry scale is essential. Zheng and co-workers in 2006 [83] reported data from fermentation at 10 m^3 scale, where the cellular astaxanthin titres reached 2.57 mg/g dry cell.

Light is capital in astaxanthin biosynthesis. Light exposure stimulates total production of carotenoids (mainly astaxanthin) in *Xanthophyllomyces* and has a negative effect on growth [72]. Therefore, the effect of white and ultraviolet light has been tested. Astaxanthin production by fermentation of *X. dendrorhous* was significantly improved at the flask scale by white light (4.0 mg/g) and by ultraviolet light (4.4 mg/g). A semi-industrial process at the 800-L scale for astaxanthin production by fermentation of *X. dendrorhous* was significantly improved by white light and glucose feeding (4.1 mg/g) [84].

4. Omics of *X. dendrorhous*: Genomics, Transcriptomics, Proteomics, and Metabolomics

As was indicated above, the metabolic redirection is a suitable reality. It can be enhanced when it is supported by omics technologies, which have been tackled in this yeast from three points of view: genomics, transcriptomics, and proteomics.

The first genomic analysis was done by using pulsed field gel electrophoresis, which defined a putative genome size of 25 Mb [85]. This genome size has been recently redefined by means of the Illumina sequencing methodology (see Table 2 for extraction protocols) [19,86]. As a result, *X. dendrorhous* presents a 19.50 Mb genome with 6385 protein-coding genes [19]. These data were slightly lower (ca. 19 Mb and 6000 ORFs) when two different strains (CBS 7918[T], CRUB 1149), in addition to the previously sequenced one (CBS 6938), were analysed. Thus, based on these three strains, which can be considered as different varieties in future, the existence of genetic heterogeneity within this red-pigmented yeast was demonstrated [87]. Genome-based pathway analysis presented sterols and carotenoids biosynthesis as the two prominent terpenoid pathways in *X. dendrorhous*. These genes do not follow the typical cluster arrangement for other specific biosynthesis pathways of fungi, which is a peculiarity of this yeast. Besides, the key regulators that control the ethanol accumulation from glucose even under aerobic conditions prior to be re-used on the stationary phase, have been provided from the genome sequence. Those are capital to reach an optimal astaxanthin production increasing the precursors availability by means of pathway engineering [19,87]. Also described was how *X. dendrorhous* is genetically fully equipped to cope with environmental oxidative stress that contributes to its genome shape [88].

The genome sequence also eased the phylogenetic analyses of this astaxanthin-producer basal agaricomycete with uncertain taxonomic placement, which presented *Wallemia* as the most basal agaricomycotinous lineage followed by Tremellomycetes. Besides, the taxonomic analysis defined a sister-group relationship between the core Tremellomycetes and the Cystofilobasidiales [19].

Two sampling points (18 and 72 h) and two different carbon sources (glucose and succinate) were used by Baeza and co-workers [86] to obtain RNA after mechanical rupture by glass beads and Tri-Reagent (Ambion, Foster City, CA, USA) extraction (Table 2). The subsequent RNAseq analysis was the base to determine sequence lengths, expression levels, GC% content, as well as the codon usage and codon context biases of the open reading frames (ORFs) [86]. A total of 1695 ORFs were the basis of the analysis in contrast with one previously described by Verdoes and van Ooyen [89] based on 10 ribosomal genes. Thus, a codon usage table of highly expressed ORFs of *X. dendrorhous* was properly defined, which is highly relevant for the heterologous expression of recombinant genes in the biotechnology industry [90].

Proteome analysis has been the most used omics methodology applied to *X. dendrorhous*, which goes from a reference map and mutant analysis to the comparison in the use of different carbon sources [91–94]. Besides, the sampling point and procedure help to discriminate those biomolecules directly involved in the process from those contaminants that include noise in the system. For example, this is the case of extracellular proteome (secretome) analysis, where proteins obtained by degradation (degradome) should be avoided [95]. To date, just the intra-cellular proteome has been analysed in *X. dendrorhous*. Those methodologies used for protein extraction, visualisation, and analysis are summarised in Table 2.

Firstly, Martínez-Moya and co-workers [93] developed a heterologous protein identification approach due to the poor genome characterisation at that time. Three time points were selected (lag, late exponential, and stationary growth phases) in minimal media (MM-glucose), where pigment accumulation is evident during the stationary phase. These authors identified two groups of up-regulated proteins: (i) carbohydrate and lipid metabolism proteins, which guarantee acetyl-CoA availability; and (ii) redox-specific proteins (e.g., monooxygenase or cytochrome P450). Both groups presented a scenario that supply the necessary redox potential for the late reactions of the astaxanthin synthesis. These results strongly suggested that astaxanthin production under aerobic conditions is a metabolic tool to scavenge ROS (reactive oxygen species) elements generated as metabolic byproducts in *X. dendrorhous* [93].

The same authors also studied by proteomics and metabolomics the influence of two different carbon sources (glucose (fermentable) or succinate (non-fermentable)) on the metabolism. Lipid and carbohydrate metabolism, carotenogenesis, as well as redox and stress responses proteins, were identified. These data confirmed the connection between the astaxanthin accumulation and oxidative stress in this yeast. Besides, the increase in acetyl-CoA availability when succinate is used as a carbon source was described, which could enhance the cellular respiration rate resulting in ROS elements that induces carotenogenesis [92].

In addition to the carbon source, the carbon to nitrogen ratio is crucial for microbial carotenoids production. This relation was analysed at fixed nitrogen concentrations by Pan and co-workers [94] in *X. dendrorhous*. Intriguingly, cell growth and astaxanthin accumulation were connected to the increasing of the carbon to nitrogen ratio, whereas the astaxanthin amount per cell was inverse. These metabolic adaptations were studied by two-dimensional electrophoresis analysis, where the up-regulation of redox- and stress-associated proteins, as well as carotenogenesis proteins were observed. In contrast, nine proteins involved in the astaxanthin synthesis were down-regulated and the de novo pigments synthesis was inhibited, which justify this peculiar unbalance between growth and cellular pigment accumulation [94].

As a result of nitrosoguanidine treatments, different coloured mutants (red, orange, pink, yellow, and white), which were analysed under the proteomics methodologies, were obtained by Barbachano-Torres and co-workers [91]. The red mutants were total carotenoids (mainly astaxanthin) overproducers, whereas orange and white ones accumulated phytoene as a result of phytoene dehydrogenase mutation. This analysis also demonstrated the close relation among the tricarboxylic acid cycle, stress response, and the carotenogenic process.

Table 2. Summary of the biomolecules extraction protocols described for X. dendrorhous (P. rhodozyma).

Media	Collection	Disruption	Analysis Method	Ref.
		DNA		
YPD medium. 21 °C, 5 days [96]	Culture: 15 mL Suspended: 0.5 mL YPD	Breaking system: 300 µL of glass beads (0.25–0.5 mm diameter). Swing mill (Retsch MM200) at a frequency of 30/s. Sample cleaning: Supernatant collected and purified by phenol/chloroform/isoamyl alcohol extraction. DNA precipitation overnight at −20 °C by 100% ice-cold ethanol (2.5 volume) and 1/10 volume of 3 M sodium acetate solution. 70% ice-cold ethanol. Dry at room temperature. DNA pellet resuspension in 30 µL H_2O. Store at 4 °C.	Agarose gel and ethidium bromide stain. Fluorescence densitometry measurement.	[19]
YM medium (100 mL). 22 °C, up to stationary phase	Centrifugation	Breaking system: DNA isolated from protoplasts: 2× wash 50 mM EDTA pH 7.5. Novozyme 234 plus LET buffer (500 mM EDTA, 7.5% 2-mercaptoethanol, 10 mM Tris pH 7.5). 16 h, 37 °C. NDS solution. (2 mg/mL proteinase K in 500 mM EDTA, 1% lauryl sarcosine and 10 mM Tris-HCl, pH 7.5). 24 h, 50 °C. Sample cleaning: Phenolic extraction (pH 8.0): 3× wash saturated phenol. 3× phenol: chloroform: isoamyl alcohol (25:24:1). 1× chloroform: isoamyl alcohol (24:1). DNA was precipitated with 98% ethanol and washed with 70% ethanol. Dry DNA resuspended in Tris: EDTA (10:1; pH 8.0) plus 40 µg/mL of RNase A. 37 °C, 30 min. Repeat phenolic extraction	DNA quantitation: 260/280 ratio (1.7–1.9) and 260/230 ratio (>2) by using a V-630 UV–vis Spectrophotometer.	[85,86]
YM broth (15 mL) at 20 °C, 72 h	Centrifugation	Breaking system: Modified phenol:chloroform:isoamyl alcohol method [97,98]: Resuspend in 500 µL lysis buffer (10 mM Tris-HCl pH 8.0, 100 mM NaCl, 1 mM EDTA (pH 8.0), 2% Triton X-100, 1% SDS). Add an equal volume of phenol/chloroform (1:1 v/v). Shake vigorously (Ika-Vibrax VXR shaker) at 1800 rpm, 20 minutes, R/T. Centrifuge at 14,000 rpm, 20 min, 4 °C. Sample cleaning: Ethanol precipitation.	N/A	[88]

Table 2. *Cont.*

Media	Collection	Disruption	Analysis Method	Ref.
		RNA		
YPD medium. 21 °C, 5 days [96]		NucleoSpin® RNA Plant kit (MACHEREY-NAGEL GmbH & Co. KG) (Following the manufacturer instructions).	RNA quality by using Nano-Photometer (IMPLEN) 1.5% agarose gel and ethidium bromide stain	[19]
Vogel minimal medium (MMv) supplemented with 2% glucose or 2% succinate	Early exponential phase (18 h) Initial stationary phase (72 h)	Breaking system: Mechanical rupture of cell pellets. 0.5 mm glass beads (BioSpec). Vortexing for 10 min. Add Tri-Reagent (Ambion). R/T 10 min. Sample cleaning: Add 200 µL of chloroform per mL of Tri-Reagent. Mix. Centrifuge: $4000\times g$, 5 min. Recover supernatant. $2\times$ acidic phenol: chloroform (1:1) extractions. Precipitate: 2 volume of isopropanol. 10 min, R/T. $1\times$ wash 75% ethanol. Resuspend in RNase-free water. RNA samples at a 260/280 ratio >1.9, measured using a V-630 UV-vis Spectrophotometer, were used for next-generation sequencing.	RNA quantitation: 260/280 ratio (>1.9) by using a V-630 UV-vis Spectrophotometer	[86]
		Proteins		
Minimal medium plus 2% glucose or succinate as carbon sources [99] Preculture: 10 mL Culture: 250 mL in 1-L flask inoculated with 2.5 mL of seed culture. 22 °C, 120 rpm	Centrifugation: $5000\times g$, 10 min, 4 °C. Pellet washed twice with ice-cold water. Centrifuge: $5000 \times g$, 10 min, 4 °C. Freeze in liquid N_2. Stored at −80 °C	Breaking system: Lyophilise cells. Add an equal volume (±500 µL) of glass beads (500 µm). Add 500 µL of lysis buffer (100 mM sodium. bicarbonate, pH 8.8, 0.5% Triton × 100, 1 mM phenylmethylsulfonyl fluoride (PMSF) and protease inhibitors). 15 min in on ice. Shake at 30 s at 4.5 m/s (RiboLyzer). Chill on ice, 1 min between shaking steps. Sample cleaning: Remove cell debris by centrifugation (15,000 rpm, 20 min, 4 °C. 10% v/v DNase-RNase solution (0.5 M Tris-HCl, pH 7.0, 0.5 M $MgCl_2$, 100 µg/mL RNAse and 2 µL DNase). 1 h, 4 °C. Add water up to 2.5 mL plus 200 µL of 0.5 M Tris (pH 6.8) and 20 µL of 1 M dithiothreitol (DTT). R/T. 30 min. 600 µL of water-saturated phenol. R/T. 30 min. Centrifuge: $5000\times g$, 10 min, 4 °C Add to supernatant 20 µL of 1 M DTT and 30 µL of 8 M ammonium acetate. R/T. 30 min. Precipitate. 2 mL of cold (−20 °C) methanol.	Bidimensional gel (pI range: 3–10 NL, 17 cm strips) Coomassie brilliant blue Trypsin digestion MALDI-TOF-MS identification	[92,93]

Table 2. Cont.

Media	Collection	Disruption	Analysis Method	Ref.
		Centrifuge: 13,000 rpm, 4 °C, 30 min. 2× wash: 70% (v/v) cold ethanol at −20 °C. Resuspend pellet: 200 μL of buffer (8 M urea, 2 M thiourea, 2% CHAPS, 0.01% [w/v] bromophenol blue). Store at −80 °C.		
YM medium 20-h old culture (beginning of carotenoid biosynthesis) 20 °C and 150 rpm.	Centrifugation: 5000 rpm, 10 min.	Breaking system: Liquid nitrogen in a mortar. Resuspend fine powder in 5 mL of buffer (50 mM Tris–HCl pH 7.4, 0.5 mM PMSF). Add 5 mL of 100 mM sodium carbonate. 1 h on ice. Centrifuge: 6000 rpm, 20 min, 4 °C. Supernatant precipitation: 10% final concentration of trichloroacetic acid (TCA). Centrifuge: 10,000 rpm for 15 min. Sample cleaning: 1× acetone. 1× 70% ethanol. 200 μL rehydration buffer (7 M urea, 2 M thiourea, 1% CHAPS, 0.5% Triton X-100, 40 mM Tris–HCl, 0.5% ampholytes 3–10, and 0.1% bromophenol blue). Centrifuge: 10,000 rpm, 15 min. Desalt: Micro Bio Spin G-30 columns (Bio-Rad) and rehydration buffer (Bio-Rad, Hercules, CA, USA).	Bidimensional gel (pI range: 3–10, 11 cm strips) Colloidal Coomassie [100]	[91]
Seed culture: 20 g/L glucose, 10 g/L yeast extract, 20 g/L peptone. 22 °C, 200 rpm, 48 h. 250-mL flask containing 30 mL. Inoculate 3 mL in a 250-mL flask containing 30 mL of seed media. 22 °C, 200 rpm, 24 h. Fermentation: (20 g/L glucose, 0.2 g/L yeast extract, 0.5 g/L $(NH_4)_2SO_4$, 1.0 g/L KH_2PO_4, 0.1 g/L NaCl, 0.5 g/L $MgSO_4 \cdot 7H_2O$, and 0.1 g/L $CaCl_2 \cdot 2H_2O$). 22 °C, 200 rpm, 144 h	Centrifugation: 8000× g, 10 min, 8 °C, 96 h.	Breaking system: Resuspend pellet in deionised water. 2× 30 kpsi high-pressure disrupter (Constant Systems Limited, Northants, UK). 10,000× g, 15 min, 4 °C. Collect supernatant 0.5 mg dissolved in 500 μL rehydration buffer [8 M urea, 4% CHAPS, 2% IPG buffer and 40 mM DTT (Dithiothreitol)].	Bidimensional gel (pI range: 4–7) Silver stain Trypsin digestion MALDI-TOF/TOF-MS identification	[94]

5. Conclusions and Future Prospects

Commercial use of astaxanthin in aquiculture, as well as in human nutrition and health is well-defined and growing year after year. Nowadays, astaxanthin production is tackled by chemical synthesis, which allows a profitable cost/benefit ratio. However, its petrochemical origin presents regulatory concerns, which is shifting attention in favour of biological sources such as bacteria, yeasts, or algae. Unfortunately, some methodological (e.g., light induction, product extraction) or biotechnological difficulties (e.g., low production) are the current key concerns of a more eco-friendly production process. The practical advances (culture conditions, transformation processes, high throughput screening systems, genetic and metabolic know-how, scale up processes) supported by omics technologies are boosting the carotenoids production by natural procedures. Thus, *X. dendrorhous*, which has been approved by the FDA for the commercial production of astaxanthin, is a good candidate to allow the natural production of astaxanthin with a proper isomerism and chemical structure. There is good knowledge and a methodological basis, but the natural production of astaxanthin has plenty of room for improvement.

Acknowledgments: We thank Josefina Merino, Bernabé Martín, Andrea Casenave and Hanna del Río for their excellent technical assistance and the doctorate and degree students of the group Laura García Calvo, Ana García-Guerra, Ana Ibáñez and Esmeralda Sastre.

Conflicts of Interest: The authors declare no conflict of interest.

References

1. Lichtenthaler, H.K.; Buschmann, C. Chlorophylls and Carotenoids: Measurement and Characterization by UV-VIS Spectroscopy. In *Current Protocols Food Analytical Chemistry*; John Wiley & Sons, Inc.: Franklin, NJ, USA, 2001. [CrossRef]
2. Sandmann, G.; Misawa, N. Fungal Carotenoids. In *Industrial Application*; Osiewacz, H.D., Ed.; Springer: Berlin/Heidelberg, Germany, 2002; pp. 247–262. [CrossRef]
3. Sieiro, C.; Poza, M.; de Miguel, T.; Villa, T.G. Genetic basis of microbial carotenogenesis. *Int. Microbiol.* **2003**, *6*, 11–16. [PubMed]
4. Fraser, P.D.; Bramley, P.M. The biosynthesis and nutritional uses of carotenoids. *Prog. Lipid Res.* **2004**, *43*, 228–265. [CrossRef] [PubMed]
5. Britton, G.; Liaaen-Jensen, S.; Pfander, H. *Carotenoids*; Birkhäuser Basel: Basel, Switzerland, 2004. [CrossRef]
6. Bhosale, P.; Bernstein, P.S. Microbial xanthophylls. *Appl. Microbiol. Biotechnol.* **2005**, *68*, 445–455. [CrossRef] [PubMed]
7. Young, A.J. The photoprotective role of carotenoids in higher plants. *Physiol. Plant.* **1991**, *83*, 702–708. [CrossRef]
8. Blomhoff, R.; Blomhoff, H.K. Overview of retinoid metabolism and function. *J. Neurobiol.* **2006**, *66*, 606–630. [CrossRef] [PubMed]
9. Walter, M.H.; Strack, D. Carotenoids and their cleavage products: Biosynthesis and functions. *Nat. Prod. Rep.* **2011**, *28*, 663–692. [CrossRef] [PubMed]
10. Fiedor, J.; Burda, K. Potential role of carotenoids as antioxidants in human health and disease. *Nutrients* **2014**, *6*, 466–488. [CrossRef] [PubMed]
11. Biesalski, H.K.; Chichili, G.R.; Frank, J.; von Lintig, J.; Nohr, D. Conversion of β-carotene to retinal pigment. *Vitam. Horm.* **2007**, *75*, 117–130. [PubMed]
12. Feltl, L.; Pacakova, V.; Stulik, K.; Volka, K. Reliability of Carotenoid Analyses: A Review. *Curr. Anal. Chem.* **2005**, *1*, 93–102. [CrossRef]
13. März, U. The Global Market for Carotenoids. Available online: https://www.bccresearch.com/market-Research/food-and-Beverage/carotenoids-Global-Market-Report-fod025e.html (accessed on 13 June 2017).
14. Panis, G.; Carreon, J.R. Commercial astaxanthin production derived by green alga *Haematococcus pluvialis*: A microalgae process model and a techno-economic assessment all through production line. *Algal Res.* **2016**, *18*, 175–190. [CrossRef]

15. Schweiggert, R.M.; Carle, R. Carotenoid Production by Bacteria, Microalgae, and Fungi. In *Carotenoids Nutrition Analysis Technology*; Kaczor, A., Baranska, M., Eds.; John Wiley & Sons, Ltd: Chichester, UK, 2016; pp. 217–240.

16. Phaff, H.; Miller, M.; Yoneyama, M.; Soneda, M. A comparative study of the yeast florae associated with trees on the Japanese Islands and on the west coast of North America. In Proceedings of the 4th Internatoinal Fermentation Symposium: Fermentation Technology Today, Kyoto, Japan, 19–25 March 1972; pp. 759–774.

17. Andrewes, A.G.; Starr, M.P. (3R,3′R)-Astaxanthin from the yeast *Phaffia rhodozyma*. *Phytochemistry* **1976**, *15*, 1009–1011. [CrossRef]

18. Sandmann, G. Carotenoids of Biotechnological Importance. In *Biotechnology of Isoprenoids*; Schrader, J., Bohlmann, J., Eds.; Springer International Publishing: Cham, Switzerland, 2014; pp. 449–467. [CrossRef]

19. Sharma, R.; Gassel, S.; Steiger, S.; Xia, X.; Bauer, R.; Sandmann, G.; Thines, M. The genome of the basal agaricomycete *Xanthophyllomyces dendrorhous* provides insights into the organization of its acetyl-CoA derived pathways and the evolution of Agaricomycotina. *BMC Genom.* **2015**, *16*, 233. [CrossRef] [PubMed]

20. Visser, H.; Sandmann, G.; Verdoes, J.C. Xanthophylls in Fungi. Metabolic Engineering of the Astaxanthin Biosynthetic Pathway in *Xanthophyllomyces dendrorhous*. In *Microbial Processes Products*; Barredo, J.-L., Ed.; Humana Press: Totowa, NJ, USA, 2005; pp. 257–272.

21. Rodríguez-Sáiz, M.; Godio, R.P.; Alvarez, V.; de la Fuente, J.L.; Martín, J.F.; Barredo, J.L. The NADP-dependent glutamate dehydrogenase gene from the astaxanthin producer *Xanthophyllomyces dendrorhous*: Use of Its promoter for controlled gene expression. *Mol. Biotechnol.* **2009**, *41*, 165–172. [CrossRef] [PubMed]

22. Gassel, S.; Breitenbach, J.; Sandmann, G. Genetic engineering of the complete carotenoid pathway towards enhanced astaxanthin formation in *Xanthophyllomyces dendrorhous* starting from a high-yield mutant. *Appl. Microbiol. Biotechnol.* **2014**, *98*, 345–350. [CrossRef] [PubMed]

23. Hara, K.Y.; Morita, T.; Mochizuki, M.; Yamamoto, K.; Ogino, C.; Araki, M.; Kondo, A. Development of a multi-gene expression system in *Xanthophyllomyces dendrorhous*. *Microb. Cell Fact.* **2014**, *13*, 175. [CrossRef] [PubMed]

24. Yamane, Y.; Higashida, K.; Nakashimada, Y.; Kakizono, T.; Nishio, N. Influence of Oxygen and Glucose on Primary Metabolism and Astaxanthin Production by *Phaffia rhodozyma* in Batch and Fed-Batch Cultures: Kinetic and Stoichiometric Analysis. *Appl. Environ. Microbiol.* **1997**, *63*, 4471–4478. [PubMed]

25. Alvarez, V.; Rodríguez-Sáiz, M.; de la Fuente, J.L.; Gudiña, E.J.; Godio, R.P.; Martín, J.F.; Barredo, J.L. The crtS gene of *Xanthophyllomyces dendrorhous* encodes a novel cytochrome-P450 hydroxylase involved in the conversion of β-carotene into astaxanthin and other xanthophylls. *Fungal Genet. Biol.* **2006**, *43*, 261–272. [CrossRef] [PubMed]

26. Ojima, K.; Breitenbach, J.; Visser, H.; Setoguchi, Y.; Tabata, K.; Hoshino, T.; van den Berg, J.; Sandmann, G. Cloning of the astaxanthin synthase gene from *Xanthophyllomyces dendrorhous* (*Phaffia rhodozyma*) and its assignment as a β-carotene 3-hydroxylase/4-ketolase. *Mol. Genet. Genom.* **2006**, *275*, 148–158. [CrossRef] [PubMed]

27. Green, A.S.; Fascetti, A.J. Meeting the Vitamin A Requirement: The Efficacy and Importance of β-Carotene in Animal Species. *Sci. World J.* **2016**, *2016*, 7393620. [CrossRef] [PubMed]

28. Chichili, G.R.; Nohr, D.; Schäffer, M.; von Lintig, J.; Biesalski, H.K. β-Carotene conversion into vitamin A in human retinal pigment epithelial cells. *Investig. Ophthalmol. Vis. Sci.* **2005**, *46*, 3562–3569. [CrossRef] [PubMed]

29. Zhang, W.; Zhang, K.Y.; Ding, X.M.; Bai, S.P.; Hernandez, J.M.; Yao, B.; Zhu, Q. Influence of canthaxanthin on broiler breeder reproduction, chick quality, and performance. *Poult. Sci.* **2011**, *90*, 1516–1522. [CrossRef] [PubMed]

30. Bone, R.A.; Landrum, J.T.; Friedes, L.M.; Gomez, C.M.; Kilburn, M.D.; Menendez, E.; Vidal, I.; Wang, W. Distribution of lutein and zeaxanthin stereoisomers in the human retina. *Exp. Eye Res.* **1997**, *64*, 211–218. [CrossRef] [PubMed]

31. Landrum, J.T.; Bone, R.A. Lutein, zeaxanthin, and the macular pigment. *Arch. Biochem. Biophys.* **2001**, *385*, 28–40. [CrossRef] [PubMed]

32. Lima, V.C.; Rosen, R.B.; Farah, M. Macular pigment in retinal health and disease. *Int. J. Retina Vitr.* **2016**, *2*, 19. [CrossRef] [PubMed]

33. Research and Markets, Global Astaxanthin Market-Sources, Technologies and Application. Available online: http://www.researchandmarkets.com/reports/3129287/global-astaxanthin-market-sources-technologies (accessed on 21 May 2017).
34. Schmidt, I.; Schewe, H.; Gassel, S.; Jin, C.; Buckingham, J.; Hümbelin, M.; Sandmann, G.; Schrader, J. Biotechnological production of astaxanthin with *Phaffia rhodozyma*/*Xanthophyllomyces dendrorhous*. *Appl. Microbiol. Biotechnol.* **2011**, *89*, 555–571. [CrossRef] [PubMed]
35. Breithaupt, D.R. Xanthophylls in Poultry Feeding. In *Carotenoids*; Britton, G., Liaaen-Jensen, S., Pfander, H., Eds.; Birkhäuser Basel: Basel, Switzerland, 2008; Volume 4, pp. 255–264.
36. Higuera-Ciapara, I.; Félix-Valenzuela, L.; Goycoolea, F.M. Astaxanthin: A review of its chemistry and applications. *Crit. Rev. Food Sci. Nutr.* **2006**, *46*, 185–196. [CrossRef] [PubMed]
37. Wang, X.; Willén, R.; Wadström, T. Astaxanthin-rich algal meal and vitamin C inhibit *Helicobacter pylori* infection in BALB/cA mice. *Antimicrob. Agents Chemother.* **2000**, *44*, 2452–2457. [CrossRef] [PubMed]
38. Park, J.S.; Chyun, J.H.; Kim, Y.K.; Line, L.L.; Chew, B.P. Astaxanthin decreased oxidative stress and inflammation and enhanced immune response in humans. *Nutr. Metab.* **2010**, *7*, 18. [CrossRef] [PubMed]
39. Yasui, Y.; Hosokawa, M.; Mikami, N.; Miyashita, K.; Tanaka, T. Dietary astaxanthin inhibits colitis and colitis-associated colon carcinogenesis in mice via modulation of the inflammatory cytokines. *Chem. Biol. Interact.* **2011**, *193*, 79–87. [CrossRef] [PubMed]
40. Fassett, R.G.; Coombes, J.S. Astaxanthin, oxidative stress, inflammation and cardiovascular disease. *Future Cardiol.* **2009**, *5*, 333–342. [CrossRef] [PubMed]
41. Li, J.; Zhu, D.; Niu, J.; Shen, S.; Wang, G. An economic assessment of astaxanthin production by large scale cultivation of *Haematococcus pluvialis*. *Biotechnol. Adv.* **2011**, *29*, 568–574. [CrossRef] [PubMed]
42. An, G.H.; Cho, M.H.; Johnson, E.A. Monocyclic carotenoid biosynthetic pathway in the yeast *Phaffia rhodozyma* (*Xanthophyllomyces dendrorhous*). *J. Biosci. Bioeng.* **1999**, *88*, 189–193. [CrossRef]
43. Rodríguez-Sáiz, M.; de la Fuente, J.L.; Barredo, J.L. *Xanthophyllomyces dendrorhous* for the industrial production of astaxanthin. *Appl. Microbiol. Biotechnol.* **2010**, *88*, 645–658. [CrossRef] [PubMed]
44. Priya, R.; Hridya, H.; Soundarya, C.; Somasundari, G.; Doss, C.G.P.; Sneha, P.; Rajasekaran, C.; Christopher, G.; Siva, R. Astaxanthin biosynthetic pathway: Molecular phylogenies and evolutionary behaviour of Crt genes in eubacteria. *Plant Gene* **2016**, *8*, 32–41. [CrossRef]
45. Andrewes, A.G.; Phaff, H.J.; Starr, M.P. Carotenoids of *Phaffia rhodozyma*, a red-pigmented fermenting yeast. *Phytochemistry* **1976**, *15*, 1003–1007. [CrossRef]
46. Sanpietro, L.M.D.; Kula, M.R. Studies of astaxanthin biosynthesis in *Xanthophyllomyces dendrorhous* (*Phaffia rhodozyma*). Effect of inhibitors and low temperature. *Yeast* **1998**, *14*, 1007–1016. [CrossRef]
47. Verdoes, J.C.; Sandmann, G.; Visser, H.; Diaz, M.; van Mossel, M.; van Ooyen, A.J.J. Metabolic engineering of the carotenoid biosynthetic pathway in the yeast *Xanthophyllomyces dendrorhous* (*Phaffia rhodozyma*). *Appl. Environ. Microbiol.* **2003**, *69*, 3728–3738. [CrossRef] [PubMed]
48. Visser, H.; van Ooyen, A.J.J.; Verdoes, J.C. Metabolic engineering of the astaxanthin-biosynthetic pathway of *Xanthophyllomyces dendrorhous*. *FEMS Yeast Res.* **2003**, *4*, 221–231. [CrossRef]
49. Liang, P.-H.; Ko, T.-P.; Wang, A.H.-J. Structure, mechanism and function of prenyltransferases. *Eur. J. Biochem.* **2002**, *269*, 3339–3354. [CrossRef] [PubMed]
50. Lee, P.C.; Schmidt-Dannert, C. Metabolic engineering towards biotechnological production of carotenoids in microorganisms. *Appl. Microbiol. Biotechnol.* **2002**, *60*, 1–11. [PubMed]
51. Misawa, N. Pathway engineering for functional isoprenoids. *Curr. Opin. Biotechnol.* **2011**, *22*, 627–633. [CrossRef] [PubMed]
52. Kajiwara, S.; Fraser, P.D.; Kondo, K.; Misawa, N. Expression of an exogenous isopentenyl diphosphate isomerase gene enhances isoprenoid biosynthesis in *Escherichia coli*. *Biochem. J.* **1997**, *324*, 421–426. [CrossRef] [PubMed]
53. Britton, G. Carotenoid Biosynthesis—An Overview. In *Carotenoids Chemistry Biology*; Krinsky, N.I., Mathews-Roth, M.M., Taylor, R.F., Eds.; Springer: Boston, MA, USA, 1989; pp. 167–184.
54. Niklitschek, M.; Alcaíno, J.; Barahona, S.; Sepúlveda, D.; Lozano, C.; Carmona, M.; Marcoleta, A.; Martínez, C.; Lodato, P.; Baeza, M.; et al. Genomic organization of the structural genes controlling the astaxanthin biosynthesis pathway of *Xanthophyllomyces dendrorhous*. *Biol. Res.* **2008**, *41*, 93–108. [CrossRef] [PubMed]

55. Verdoes, J.C.; Krubasik, K.P.; Sandmann, G.; van Ooyen, A.J. Isolation and functional characterisation of a novel type of carotenoid biosynthetic gene from *Xanthophyllomyces dendrorhous*. *Mol. Gen. Genet.* **1999**, *262*, 453–461. [CrossRef] [PubMed]

56. Ajikumar, P.K.; Tyo, K.; Carlsen, S.; Mucha, O.; Phon, T.H.; Stephanopoulos, G. Terpenoids: Opportunities for biosynthesis of natural product drugs using engineered microorganisms. *Mol. Pharm.* **2008**, *5*, 167–190. [CrossRef] [PubMed]

57. Verdoes, J.C.; Misawa, N.; van Ooyen, A.J. Cloning and characterization of the astaxanthin biosynthetic gene encoding phytoene desaturase of *Xanthophyllomyces dendrorhous*. *Biotechnol. Bioeng.* **1999**, *63*, 750–755. [CrossRef]

58. Krubasik, P.; Sandmann, G. A carotenogenic gene cluster from *Brevibacterium linens* with novel lycopene cyclase genes involved in the synthesis of aromatic carotenoids. *Mol. Gen. Genet.* **2000**, *263*, 423–432. [CrossRef] [PubMed]

59. Alcaíno, J.; Barahona, S.; Carmona, M.; Lozano, C.; Marcoleta, A.; Niklitschek, M.; Sepúlveda, D.; Baeza, M.; Cifuentes, V. Cloning of the cytochrome p450 reductase (*crtR*) gene and its involvement in the astaxanthin biosynthesis of *Xanthophyllomyces dendrorhous*. *BMC Microbiol.* **2008**, *8*, 169. [CrossRef] [PubMed]

60. Ukibe, K.; Hashida, K.; Yoshida, N.; Takagi, H. Metabolic engineering of *Saccharomyces cerevisiae* for astaxanthin production and oxidative stress tolerance. *Appl. Environ. Microbiol.* **2009**, *75*, 7205–7211. [CrossRef] [PubMed]

61. Calo, P.; González, T. The yeast *Phaffia rhodozyma* as an industrial source of astaxanthin. *Microbiologia* **1995**, *11*, 386–388. [PubMed]

62. Alcaino, J.; Baeza, M.; Cifuentes, V. Astaxanthin and Related Xanthophylls. In *Biosynthesis Molecular Genetics of Fungal Secondary Metabolites*; Martín, J.-F., Garcia-Estrada, C., Zeilinger, S., Eds.; Springer: New York, NY, USA, 2014; pp. 187–207.

63. Ukibe, K.; Katsuragi, T.; Tani, Y.; Takagi, H. Efficient screening for astaxanthin-overproducing mutants of the yeast *Xanthophyllomyces dendrorhous* by flow cytometry. *FEMS Microbiol. Lett.* **2008**, *286*, 241–248. [CrossRef] [PubMed]

64. An, G.H.; Schuman, D.B.; Johnson, E.A. Isolation of *Phaffia rhodozyma* Mutants with Increased Astaxanthin Content. *Appl. Environ. Microbiol.* **1989**, *55*, 116–124. [PubMed]

65. Retamales, P.; León, R.; Martínez, C.; Hermosilla, G.; Pincheira, G.; Cifuentes, V. Complementation analysis with new genetic markers in *Phaffia rhodozyma*. *Antonie Van Leeuwenhoek* **1998**, *73*, 229–236. [CrossRef] [PubMed]

66. De la Fuente Moreno, J.L.; Peiro, E.; Díez García, B.; Marcos Rodríguez, A.T.; Schleissner, C.; Rodríguez Saiz, M.; Rodríguez-Otero, C.; Cabri, W.; Barredo, J.L. Method of Producing Astaxanthin by Fermenting Selected Strains of *Xanthophyllomyces dendrorhous*. Patent EP1479777A1, 3 February 2003.

67. Liu, Z.Q.; Zhang, J.F.; Zheng, Y.G.; Shen, Y.C. Improvement of astaxanthin production by a newly isolated Phaffia rhodozyma mutant with low-energy ion beam implantation. *J. Appl. Microbiol.* **2008**, *104*, 861–872. [CrossRef] [PubMed]

68. Ni, H.; Hong, Q.; Xiao, A.; Li, L.; Cai, H.; Su, W. Characterization and evaluation of an astaxanthin over-producing *Phaffia rhodozyma*. *Sheng Wu Gong Cheng Xue Bao (Chinese J. Biotechnol.)* **2011**, *27*, 1065–1075.

69. Johnson, E.A. *Phaffia rhodozyma*: Colorful odyssey. *Int. Microbiol.* **2003**, *6*, 169–174. [CrossRef] [PubMed]

70. Chumpolkulwong, N.; Kakizono, T.; Nagai, S.; Nishio, N. Increased astaxanthin production by *Phaffia rhodozyma* mutants isolated as resistant to diphenylamine. *J. Ferment. Bioeng.* **1997**, *83*, 429–434. [CrossRef]

71. Schroeder, W.A.; Calo, P.; DeClercq, M.L.; Johnson, E.A. Selection for carotenogenesis in the yeast *Phaffia rhodozyma* by dark-generated singlet oxygen. *Microbiology* **1996**, *142*, 2923–2929. [CrossRef]

72. An, G.H.; Johnson, E.A. Influence of light on growth and pigmentation of the yeast *Phaffia rhodozyma*. *Antonie Van Leeuwenhoek* **1990**, *57*, 191–203. [CrossRef] [PubMed]

73. Pérez-García, F.; Vasco-Cárdenas, M.F.; Barreiro, C. Biotypes analysis of *Corynebacterium glutamicum* growing in dicarboxylic acids demonstrates the existence of industrially-relevant intra-species variations. *J. Proteom.* **2016**, *146*, 172–183. [CrossRef] [PubMed]

74. Breitenbach, J.; Visser, H.; Verdoes, J.C.; van Ooyen, A.J.J.; Sandmann, G. Engineering of geranylgeranyl pyrophosphate synthase levels and physiological conditions for enhanced carotenoid and astaxanthin synthesis in *Xanthophyllomyces dendrorhous*. *Biotechnol. Lett.* **2011**, *33*, 755–761. [CrossRef] [PubMed]

75. Gassel, S.; Schewe, H.; Schmidt, I.; Schrader, J.; Sandmann, G. Multiple improvement of astaxanthin biosynthesis in *Xanthophyllomyces dendrorhous* by a combination of conventional mutagenesis and metabolic pathway engineering. *Biotechnol. Lett.* **2013**, *35*, 565–569. [CrossRef] [PubMed]

76. Wery, J.; Verdoes, J.C.; van Ooyen, A.J.J. Efficient Transformation of the Astaxanthin-Producing Yeast *Phaffia rhodozyma*. *Biotechnol. Tech.* **1998**, *12*, 399–405. [CrossRef]

77. Yamamoto, K.; Hara, K.Y.; Morita, T.; Nishimura, A.; Sasaki, D.; Ishii, J.; Ogino, C.; Kizaki, N.; Kondo, A. Enhancement of astaxanthin production in *Xanthophyllomyces dendrorhous* by efficient method for the complete deletion of genes. *Microb. Cell Fact.* **2016**, *15*, 155. [CrossRef] [PubMed]

78. Hara, K.Y.; Morita, T.; Endo, Y.; Mochizuki, M.; Araki, M.; Kondo, A. Evaluation and screening of efficient promoters to improve astaxanthin production in *Xanthophyllomyces dendrorhous*. *Appl. Microbiol. Biotechnol.* **2014**, *98*, 6787–6793. [CrossRef] [PubMed]

79. Martín, J.F.; Gudiña, E.; Barredo, J.L. Conversion of β-carotene into astaxanthin: Two separate enzymes or a bifunctional hydroxylase-ketolase protein? *Microb. Cell Fact.* **2008**, *7*, 3. [CrossRef] [PubMed]

80. Lodato, P.; Alcaíno, J.; Barahona, S.; Niklitschek, M.; Carmona, M.; Wozniak, A.; Baeza, M.; Jiménez, A.; Cifuentes, V. Expression of the carotenoid biosynthesis genes in *Xanthophyllomyces dendrorhous*. *Biol. Res.* **2007**, *40*, 73–84. [CrossRef] [PubMed]

81. Chi, S.; He, Y.; Ren, J.; Su, Q.; Liu, X.; Chen, Z.; Wang, M.; Li, Y.; Li, J. Overexpression of a bifunctional enzyme, CrtS, enhances astaxanthin synthesis through two pathways in *Phaffia rhodozyma*. *Microb. Cell Fact.* **2015**, *14*, 90. [CrossRef] [PubMed]

82. Ledetzky, N.; Osawa, A.; Iki, K.; Pollmann, H.; Gassel, S.; Breitenbach, J.; Shindo, K.; Sandmann, G. Multiple transformation with the *crtYB* gene of the limiting enzyme increased carotenoid synthesis and generated novel derivatives in *Xanthophyllomyces dendrorhous*. *Arch. Biochem. Biophys.* **2014**, *545*, 141–147. [CrossRef] [PubMed]

83. Zheng, Y.G.; Hu, Z.C.; Wang, Z.; Shen, Y.C. Large-Scale Production of Astaxanthin by *Xanthophyllomyces dendrorhous*. *Food Bioprod. Process.* **2006**, *84*, 164–166. [CrossRef]

84. De la Fuente, J.L.; Rodríguez-Sáiz, M.; Schleissner, C.; Díez, B.; Peiro, E.; Barredo, J.L. High-titer production of astaxanthin by the semi-industrial fermentation of *Xanthophyllomyces dendrorhous*. *J. Biotechnol.* **2010**, *148*, 144–146. [CrossRef] [PubMed]

85. Cifuentes, V.; Hermosilla, G.; Martínez, C.; León, R.; Pincheira, G.; Jiménez, A. Genetics and electrophoretic karyotyping of wild-type and astaxanthin mutant strains of *Phaffia rhodozyma*. *Antonie Van Leeuwenhoek* **1997**, *72*, 111–117. [CrossRef] [PubMed]

86. Baeza, M.; Alcaíno, J.; Barahona, S.; Sepúlveda, D.; Cifuentes, V. Codon usage and codon context bias in *Xanthophyllomyces dendrorhous*. *BMC Genom.* **2015**, *16*, 1–12. [CrossRef] [PubMed]

87. Wozniak, A.; Lozano, C.; Barahona, S.; Niklitschek, M.; Marcoleta, A.; Alcaíno, J.; Sepulveda, D.; Baeza, M.; Cifuentes, V. Differential carotenoid production and gene expression in *Xanthophyllomyces dendrorhous* grown in a nonfermentable carbon source. *FEMS Yeast Res.* **2011**, *11*, 252–262. [CrossRef] [PubMed]

88. Bellora, N.; Moliné, M.; David-Palma, M.; Coelho, M.A.; Hittinger, C.T.; Sampaio, J.P.; Gonçalves, P.; Libkind, D. Comparative genomics provides new insights into the diversity, physiology, and sexuality of the only industrially exploited tremellomycete: *Phaffia rhodozyma*. *BMC Genom.* **2016**, *17*, 901. [CrossRef] [PubMed]

89. Verdoes, J.C.; van Ooyen, A.J.J. Codon usage in *Xanthophyllomyces dendrorhous* (formerly *Phaffia rhodozyma*). *Biotechnol. Lett.* **2000**, *22*, 9–13. [CrossRef]

90. Elena, C.; Ravasi, P.; Castelli, M.E.; Peirú, S.; Menzella, H.G. Expression of codon optimized genes in microbial systems: Current industrial applications and perspectives. *Front. Microbiol.* **2014**, *5*, 21. [CrossRef] [PubMed]

91. Barbachano-Torres, A.; Castelblanco-Matiz, L.M.; Ramos-Valdivia, A.C.; Cerda-García-Rojas, C.M.; Salgado, L.M.; Flores-Ortiz, C.M.; Ponce-Noyola, T. Analysis of proteomic changes in colored mutants of Xanthophyllomyces dendrorhous (*Phaffia rhodozyma*). *Arch. Microbiol.* **2014**, *196*, 411–421. [CrossRef] [PubMed]

92. Martinez-Moya, P.; Niehaus, K.; Alcaíno, J.; Baeza, M.; Cifuentes, V. Proteomic and metabolomic analysis of the carotenogenic yeast *Xanthophyllomyces dendrorhous* using different carbon sources. *BMC Genom.* **2015**, *16*, 289. [CrossRef] [PubMed]

93. Martinez-Moya, P.; Watt, S.A.; Niehaus, K.; Alcaíno, J.; Baeza, M.; Cifuentes, V. Proteomic analysis of the carotenogenic yeast *Xanthophyllomyces dendrorhous*. *BMC Microbiol.* **2011**, *11*, 131. [CrossRef] [PubMed]

94. Pan, X.; Wang, B.; Gerken, H.G.; Lu, Y.; Ling, X. Proteomic analysis of astaxanthin biosynthesis in *Xanthophyllomyces dendrorhous* in response to low carbon levels. *Bioprocess Biosyst. Eng.* **2017**. [CrossRef] [PubMed]

95. Barreiro, C.; Martín, J.F.; García-Estrada, C. Proteomics Shows New Faces for the Old Penicillin *Producer Penicillium chrysogenum*. *J. Biomed. Biotechnol.* **2012**, *2012*, 1–15. [CrossRef] [PubMed]

96. Ausubel, F.; Brent, R.; Kingston, R.; Moore, D.; Seidman, J.; Smith, J.; Struhl, K. YPD media. In *Current Protocols Molecular Biology Cold Spring Harbor Protocols*; Cold Spring Harbor Laboratory Press: Brooklyn, NY, USA, 1994. [CrossRef]

97. Gonçalves, P.; Valério, E.; Correia, C.; de Almeida, J.M.G.C.F.; Sampaio, J.P. Evidence for divergent evolution of growth temperature preference in sympatric Saccharomyces species. *PLoS ONE* **2011**, *6*, e20739. [CrossRef] [PubMed]

98. Soni, R.; Murray, J.A. A rapid and inexpensive method for isolation of shuttle vector DNA from yeast for the transformation of *E. coli*. *Nucleic Acids Res.* **1992**, *20*, 5852. [CrossRef] [PubMed]

99. Retamales, P.; Hermosilla, G.; León, R.; Martínez, C.; Jiménez, A.; Cifuentes, V. Development of the sexual reproductive cycle of *Xanthophyllomyces dendrorhous*. *J. Microbiol. Methods* **2002**, *48*, 87–93. [CrossRef]

100. Candiano, G.; Bruschi, M.; Musante, L.; Santucci, L.; Ghiggeri, G.M.; Carnemolla, B.; Orecchia, P.; Zardi, L.; Righetti, P.G. Blue silver: A very sensitive colloidal Coomassie G-250 staining for proteome analysis. *Electrophoresis* **2004**, *25*, 1327–1333. [CrossRef] [PubMed]

Journal of
Fungi

MDPI

Article

Microscopic Analysis of Pigments Extracted from Spalting Fungi

Sarath M. Vega Gutierrez * and Sara C. Robinson

Department of Wood Science & Engineering, 119 Richardson Hall, Oregon State University, Corvallis, OR 97331, USA; sara.robinson@oregonstate.edu
* Correspondence: sarath.vega@oregonstate.edu; Tel.: +1-541-745-9081

Academic Editors: Laurent Dufossé, Yanis Caro and Mireille Fouillaud
Received: 2 February 2017; Accepted: 10 March 2017; Published: 14 March 2017

Abstract: Pigments that are currently available in the market usually come from synthetic sources, or, if natural, often need mordants to bind to the target substrate. Recent research on the fungal pigment extracts from *Scytalidium cuboideum*, *Scytalidium ganodermophthorum*, *Chlorociboria aeruginosa*, and *Chlorociboria aeruginascens* have been shown to successfully dye materials, like wood, bamboo, and textiles, however, there is no information about their binding mechanisms. Due to this, a microscopic study was performed to provide information to future manufacturers interested in these pigments. The results of this study show that *S. ganodermophthorum* and *C. aeruginosa* form an amorphous layer on substrates, while *S. cuboideum* forms crystal-like structures. The attachment and morphology indicate that there might be different chemical and physical interactions between the extracted pigments and the materials. This possibility can explain the high resistance of the pigments to UV light and color fastness that makes them competitive against synthetic pigments. These properties make these pigments a viable option for an industry that demands natural pigments with the properties of the synthetic ones.

Keywords: extracted fungal pigments; spalting; microscopy; SEM; FIB

1. Introduction

The impact of synthetic dyes on the textile market is significant, in terms of the dyes' ecological impact [1], and most dyes in use today come from synthetic sources. For example, the commonly used red pigments of ferrite red oxide and Venetian red contain iron oxide, cadmium, and copper oxide. One of the most common yellow pigments, "lead chromate", and the popular green pigment, "chromium oxide", contain the aforementioned elements [2]. Other commonly used dyes, such as aniline dyes, are amino based, and are a derivate from petroleum [3].

Natural pigment and dye alternatives do exist, although they are not competitive in the market anymore for a number of reasons. Many of these natural dyes come from plants, lichens, and insects, and may exhibit issues with color fastness [4], adherence [4], UV stability [5], and toxicity [6], among others, making them less competitive compared with their renewable counterparts. Studies on bacterial-produced pigments have promising results for their use in textiles [7], although work with fungal pigments has been disappointing. Historic fungal pigment work focused on the use of fruiting bodies of some fungal species for dye extractions, but these dyes had the same issues as other natural dyes (low resistance to UV light, need for mordants). Lichens, likewise, have similar issues [8]. However, recent research by Hinsch showed the potential of spalting fungi pigments for their application in textiles in terms of their color fastness, crocking, and stability in UV light, all without mordants [9]. Recent research on a specific group of spalting fungi, fungi that produce extracellular pigments into wood, has shown that these pigments have the potential to equal, and in some instances even outperform, synthetic pigments [10].

Pigment-type spalting fungi are a select group of soft-rotting ascomycetes that have been shown to reliably dye a number of substrates, including wood [11], bamboo [12], and textiles [9,13]. These pigments have been found to be light fast, color fast [9], and UV light stable [9,10,14]. The specific pigments of interest in these spalting fungi are as follows: *Scytalidium cuboideum* (Sacc. And Ellis) Singler and Kang, which produces a red pigment called draconin red [9]. *Scytalidium ganodermophthorum* Kand, Singler, Y.W., Lee and S.H., Yun, produces an unidentified yellow pigment [9,15], and *Chlorociboria aeruginosa* (Oeder) Seaver and *Chlorociboria aeruginascens* (Nyl.) Kanouse ex C.S. Ramamurthi, Korf and L.R. Batra. produce xylindein, a naphthoquinone blue-green pigment [16,17].

The commercial viability of the extracted pigments obtained from spalting fungi, especially if they are to be used within the textile and wood markets, requires a better understanding of their microscopic characteristics. Previous microscopic studies have focused on naturally-produced fungal pigments. Tudor et al. focused mainly in melanin producing spalting fungi in natural form, finding that the pigments located mainly in rays and fibers. Blanchette et al. used transmission electron microscopy (TEM) to verify the presence of the fungus *Chlorociboria* sp. in art pieces of the 1400s, finding evidence of hyphae mostly in rays and vessels of the wood of the analyzed art pieces [18], and Michaelsen et al. used thin-layer chromatography and mass spectrometry to identify the pigments produced by *Chlorociboria* in art pieces of the sixteenth to eighteenth century [19]. These studies have focused mainly on the pigments produced naturally by the fungi in wood. No previous works have been done on extracted fungal pigments from these specific spalting fungi under laboratory growth conditions, likely due to the fact that the pigment extraction is relatively recent [20]. Most of the pigment extraction research has been done on the genus *Chlorociboria* [21], but it was not until Robinson et al. developed a standard method to extract the pigments from the genera *Chlorociboria* and *Scytalidium* with dichloromethane (DCM) [20] that pigment extraction from spalting fungi became a common method for working with fungal pigments.

These extracted pigments have a wide spectrum of possible applications in the wood finish and textile fields. Before this can be accomplished, however, it will be necessary to better understand the spalting pigments: how they look, how they interact, and where they are deposited on their substrates. In this study, the pigments of *Chlorociboria* sp., *S. ganodermophthorum*, and *S. cuboideum*, were characterized in terms of deposition on wood and textile substrates. The information gathered from this microscopic study will give a broader understanding of how the extracted fungal pigments work, and whether or not they effectively deposit directly onto substrates, such as wood and textiles. This is important because there are no prior studies on how extracted fungal pigments deposit onto materials on a microscopic level. How dyes deposit is critical to set a basis for future research and to gain an understanding of their potential ability to bind to, and stay on, potential substrates.

2. Materials and Methods

Different microscopy techniques were required for the different pigments studied. All of the techniques are detailed below.

2.1. Light Microscopy

Fourteen-millimeter cubes of cottonwood (*Populus trichocarpa* Torr. and A. Gray) were treated with 60 drops of extracted pigments of *S. cuboideum* UAMH 11517 (isolated from *Quercus* sp. in Memphis, TN), *S. ganodermophthorum* UAMH 10320 (isolated from oak wood logs in Gyeonggi Province, South Korea) and *C. aeruginosa* UAMH 11657 (isolated from a decaying hardwood log in Haliburton, ON, Canada) carried in dichloromethane (DCM). All of the pigments were standardized following the values set by Robinson et al. [11,22]. The pigment drops were applied in the standard amount on the cross-section of wood blocks following the protocol of Robinson et al. [11].

After application, the blocks were air dried for 48 h in a fume hood at 20 °C to evaporate the DCM, then cut into slices between 10–14 μm with a Spencer Buffalo microtome (Spencer Lens Co.,

Buffalo, NY, USA). The wood slides were mounted on VWR (VWR, Radnor, PA, USA) glass slides and covered with VWR #1 glass cover slides. The imaging was done with a Nikon Eclipse Ni-U equipped with a Nikon DS-Ri2 camera (Nikon Instruments Inc., Melville, NY, USA). Samples were analyzed with special focus on the rays and vessels of the samples. Coordinates were taken from the areas that showed a higher wood coloration by the pigments for further analysis.

2.2. Confocal Microscopy

The samples used for the light microscopy analysis were also used for their evaluation with confocal microscopy. A Zeiss LSM 780 NLO (Carl Zeiss Microscopy, LLC, Thornwood, NY, USA) was used for this analysis. The coordinates taken during the light microscopy analysis were used on this experiment, as they were the areas of interest. Each area of interest was analyzed with a set power of the argon laser (488 nm) at 0%–2%, and the filters were set from 405 to 633 nm. The wider use of filters was used to identify the optimum combination to try to distinguish the pigments from the wood. The images were obtained with the Zeiss ZEN software and analyzed with the Zeiss ZEN 2.3 lite freeware (Carl Zeiss Microscopy).

2.3. Scanning Electron Microscopy (SEM)

Silica squares (2 cm × 0.5 mm) were used as a base to apply extracted pigments as a control. Fifteen drops of each fungal pigment, solubilized in DCM at a standardized concentration were applied on them [11,22]. The DCM was allowed to evaporate between each drop. Each silica piece was fixed to a Ted Pella, Inc. (Ted Pella, Inc., Redding, CA, USA) aluminum stud of one centimeter in diameter for SEM. Ted Pella, Inc. double-coated carbon conductive tape was used to fix the samples to the aluminum studs.

Wood pieces of sugar maple and cottonwood (6 mm × 3 mm × 0.3 mm) were prepared, with the wider face oriented on the radial cut. Three pieces of each were treated on a similar way than the wood blocks, using 10 and 40 drops of each extracted fungal pigment, allowing the DCM to evaporate between drops.

Fabric squares of cotton and polyester (Testfabrics, Inc., West Pittston, PA, USA) of 5 mm × 5 mm were treated with 40 drops of extracted pigment of *S. cuboideum*, letting the DCM carrier to evaporate between drop applications. The selection of the pigment of *S. cuboideum* was done considering its characteristic morphology that would allow an easier identification of the pigments and the fibers.

The samples were mounted on Ted Pella, Inc. aluminum studs of one centimeter in diameter for SEM. A Ted Pella, Inc. double-coated carbon conductive tape was used to fix the samples to the studs. A sputter coating of gold-palladium was applied with a Cressington Sputter Coater 108 Auto (Cressington Scientific Instruments, Inc., Cranberry Twp, PA, USA) for giving the samples an enhanced optical contrast on all of the samples. The samples were exposed for 35 s to develop an even coating of 30–45 nm to avoid the electron charging of the samples and an enhanced optical contrast.

The samples were placed in the stage of an FEI QUANTA 600F environmental SEM (FEI Co., Hillsboro, OR, USA). The samples were viewed at an electron spot size of 4.5 to 5, and a high voltage (HV) between 10 to 20 kV.

2.4. Energy-Dispersive X-Ray Spectroscopy (EDS)

The extracted fungal pigment control samples were analyzed using the software EDAX Genesis (EDAX Inc., Mahwah, NJ, USA) for EDS analysis. For the analysis different sections of the samples were selected in the SEM and they were read using the X-ray detector. The software was calibrated for the detection of basic organic and inorganic elements. The limitation of the analysis is that it cannot detect hydrogen (H) due to its low molecular weight. For this reason, the analysis was focused on the detection of carbon (C), oxygen (O) and other elements that could have been present in the samples.

2.5. Focused Ion Beam (FIB)

The samples that showed higher saturation of the pigments on their surface during the SEM analysis were selected for performing an FIB cut. The selected samples were sputter coated with 10 nm of chromium oxide (Cr_2O_3) using a Varian Vacuum Evaporator VE10 (Agilent Technologies, Santa Clara, CA, USA). For the sputter coating, the bell of the sputter coater was opened and pieces of chromium (Cr) were placed in the holder connected to the back electrodes. On the stage below the holder, the aluminum studs with the wood samples were placed. Then the bell was closed and the vacuum system started. The first vacuum value required was 4×10^{-1} kPa, after reaching this value; the high-speed vacuum was activated to reach a pressure of 1×10^{-3} kPa. After obtaining the desired vacuum pressure, the stage where the samples were placed was set to a rotating speed of 10 rpm. After setting the vacuum and stage rotation, the electrodes where the holder with the Cr was connected was soaked at a power level of 2 for 60 s. Then, the power was increased to level 6 in 10 s. At this power level, the vaporization of the Cr and formation of the Cr_2O_3 started. The system remained at this level for 30 s to obtain the desired coating thickness. After 30 s, the system was turned off and the vacuum system was allowed to reach atmospheric pressure before being opened to retrieve the samples.

The coated samples were then placed in an FEI QUANTA 3D dual beam SEM/FIB (FEI Co., Hillsboro, OR, USA). Imaging of the samples was done to identify areas with higher saturation of pigments. On these areas, a rectangle of 20 μm × 3 μm was outlined. On this rectangle, two coating deposits were applied. The first coating was a layer of 20 μm × 3 μm × 100 nm of carbon (C). The second layer of coating was platinum (Pt) with the same size and depth as the carbon one. Both deposits were applied at a current of 5 Kv with 3.4 nA. After applying the deposits, the ion beam was turned on and the stage was tilted 52 degrees. The coated area was found with the ion beam and focused. There, a deposition of 1 μm of C and Pt was applied with a current of 5 Kv and 3.4 nA to protect the pre-coated area from the ion beam. Then the ion beam was activated to create a laddered cross-section of 15 μm × 1.5 μm × 5 μm at 5 Kv and 3 nA on the wood samples. The depth was enough to erode the protective coating layers, the pigment and the wood cell wall. After finishing the cross-section cutting, the borders were cleaned with the ion beam at a depth of 1 μm, to enhance the surface and clean debris generated during the cutting.

3. Results

3.1. Light Microscopy

The samples analyzed with light microscopy showed the pigments tended to accumulate in the wood vessels. Samples treated with *S. cuboideum* pigment extract showed filament-like structures accumulated on the vessels, especially in the helical thickenings of cottonwood, as shown in Figure 1.

The wood slides that contained *C. aeruginosa* pigments tended to accumulate in the intervessel pits (see Figure 2). Compared with the pigments from *S. cuboideum*, *C. aeruginosa* pigment formed what looks like an even layer on the vessel wall and it also covered the helical thickenings.

With light microscopy, the pigment of *S. ganodermophthorum* was visible on the vessel cell walls, and it also tended to accumulate on the helical thickenings (see Figure 3). Compared with the other two pigments, the yellow pigment tended to create a thicker layer on the vessel cell walls in contrast with the pigment from *C. aeruginosa*. It also accumulated between the helical thickenings, but compared to *S. cuboideum*, the pigment of *S. ganodermophthorum* did not show a characteristic shape.

Figure 1. Vessel cell from cottonwood with extracted pigments of *S. cuboideum* between the helical thickenings, pointed to with arrows. Picture taken with a Nikon Eclipse Ni-U at a magnification of 20×.

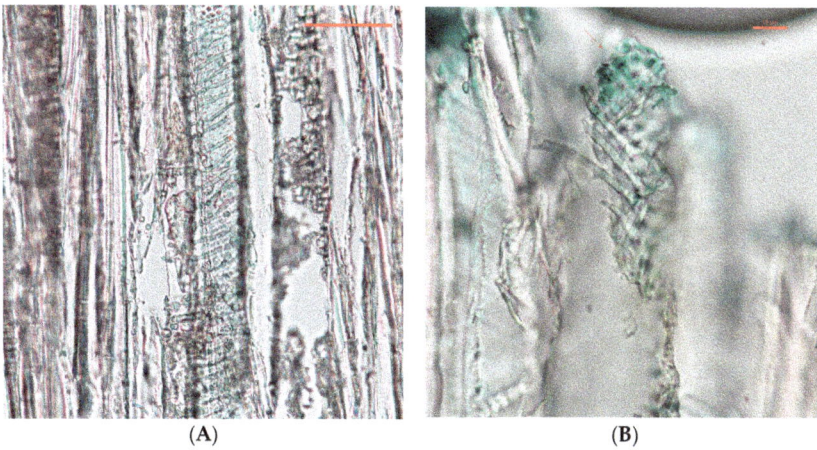

(A)   (B)

Figure 2. (**A**) Vessel cell wall of cottonwood covered with extracted pigment, pointed to with arrows, from *C. aeruginosa* taken at a 20× magnification; (**B**) Detail of the vessel cell wall showing concentration of the pigment in the vessel pits at a 40× magnification. Picture taken with a Nikon Eclipse Ni-U.

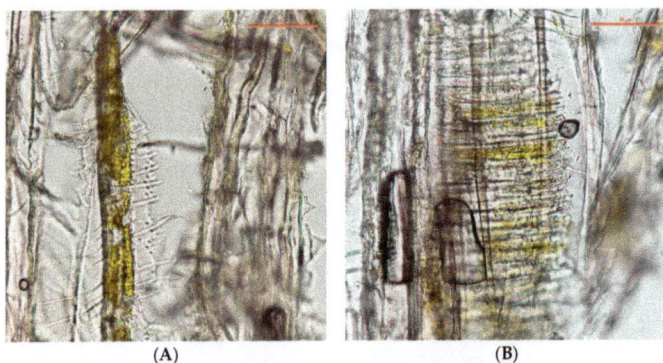

Figure 3. (**A**) Vessels of cottonwood with extracted pigment of *S. ganodermophthorum* accumulating on the cell walls; (**B**) Pigments of *S. ganodermophthorum* between the helical thickenings, pointed to with arrows. Picture taken with a Nikon Eclipse Ni-U at 20× magnification.

3.2. Confocal Microscopy

With confocal microscopy, no difference between pigments and wood was visible due to the autofluorescence of wood and the broad-spectrum fluorescence of the extracted pigments at the wavelengths of 405, 488 and 561 nm. Images of the pure dry extracted pigments showed that they had textures. The red pigment had some peaks visible when analyzed at 561 nm. The yellow pigment showed an uneven texture when analyzed at 488 nm. The green pigment imaged at 488 nm showed an almost even surface. However, in general, this method showed to not be adequate for further analysis.

3.3. SEM on Dry Extracted Fungal Pigments

Results with SEM were the most promising. Silica squares with pure *S. cuboideum* pigment showed that they crystalize and form flower-like structures, as shown in Figure 4. This structure is composed by filament-like elements, which are consistent with the results obtained with light and confocal microscopy.

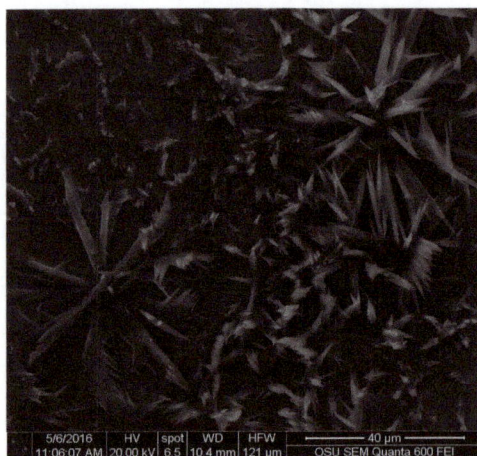

Figure 4. Crystal like structures from the extracted pigment of *S. cuboideum*. Picture taken with an FEI QUANTA 600 F.

The controls for *C. aeruginosa*, showed that the pigment has an amorphous structure, but it also showed two distinct areas, which can correspond to a molecular differences that affects the way that the electrons impact the compound (see Figure 5). Additionally, this pigment does not show a characteristic shape as the pigment from *S. cuboideum*, but it resembles a film.

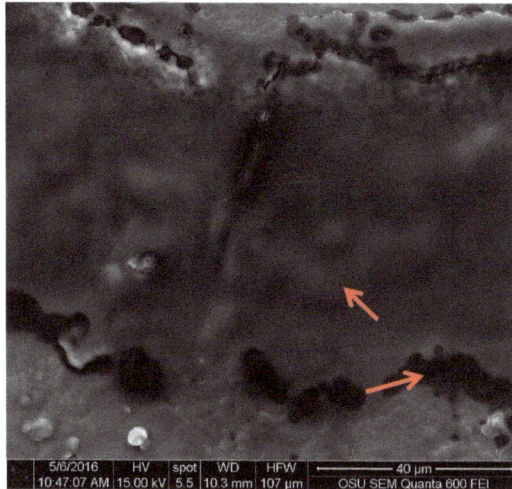

Figure 5. Extracted pigment of *C. aeruginosa* showing two reflective electron areas, pointed to with arrows. Picture taken with an FEI QUANTA 600 F.

The two electron reflective surface was also observed for *S. ganodermophthorum* (see Figure 6). The control also showed a more rugged surface, compared to the one of *C. aeruginosa*. This uneven surface was also observed with the light microscope and confocal microscopy, but with the use of SEM it is possible to have a more detailed view of the topography of this pigment.

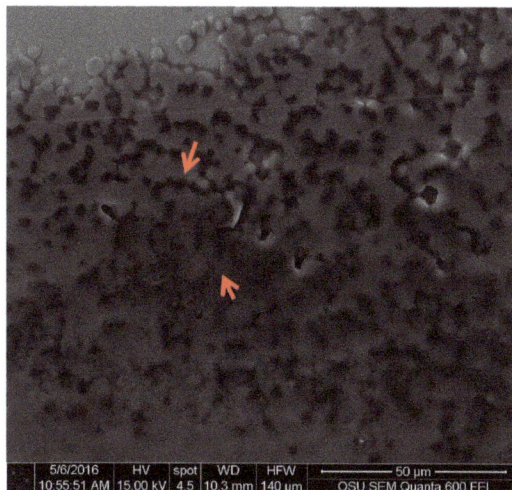

Figure 6. Extracted pigment of *S. ganodermophthorum* showing texture. Different electron reflectance is indicated with arrows. Picture taken with an FEI QUANTA 600 F.

3.4. SEM on Wood

For the wood samples, all three pigments were deposited on the surface of wood. For *S. cuboideum*, the crystal-like structures accumulated on top of the cell walls on the radial section, and on the extremes of the cell walls. They also tended to group as seen the dry pigment on silica where the pigment forms the flower-like structure (see Figure 7) that could reach up to an average of 81 μm of diameter.

Areas with less concentration of pigment presented smaller crystals on the surface of the wood cell walls shows small crystals forming on the surface of a vessel cell wall.

For *C. aeruginosa* and *S. ganodermophthorum*, the pigments formed an uneven film, covering the surface of all of the cell elements of the wood samples (see Figure 8).

Figure 7. Flower-like structures formed by the extracted pigment of *S. cuboideum* on the fiber cells of cottonwood. Picture taken with an FEI QUANTA 600 F.

Figure 8. Amorphous layer formed by the extracted pigment of *C. aeruginosa*. Picture taken with an FEI QUANTA 600 F.

3.5. SEM on Textiles

The fabrics showed that the pigment of *S. cuboideum* formed crystal-like structures that wrapped to the fiber surface of the polyester sample as shown on Figure 9.

Figure 9. Pigment of *S. cuboideum* wrapping around polyester fibers. Picture taken with an FEI QUANTA 600 F.

The behavior of the pigment on cotton was similar to wood, tending to accumulate at the ends of the fibers and the pigment tended to form convoluted areas on top of the cotton fibers as shown in Figure 10.

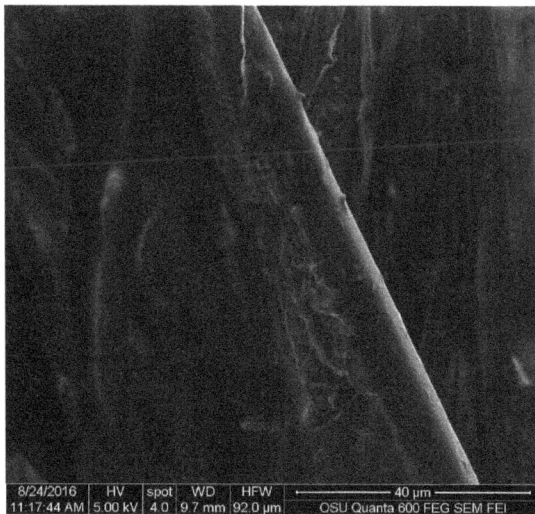

Figure 10. Extracted pigment of *S. cuboideum* accumulating on top of a cotton fiber. Picture taken with an FEI QUANTA 600 F.

3.6. EDX Analysis

The EDX analysis showed that the elements contained by the three pigments were carbon (C) and oxygen (O). No data were recorded for hydrogen (H) due to the limitation of the machine to detect that element. This made the calculation of the elements show up to 100% between C and O. In general, the pigments showed similar amounts of C and O in their structure, and we can infer the presence of hydrogen (H) on the three of them (see Table 1).

Table 1. Percentage of carbon (C) and oxygen (O) present in the extracted fungal pigments.

Extracted Pigment	Carbon (C) %	Oxygen (O) %	Silica (Si) and Gold (Au) % from Coating and Substrate	Total
Draconin red from *Scytalidium cuboideum*	78%	20%	2%	100%
Yellow pigment from *Scytalidium ganodermophthorum*	79%	21%	0%	100%
Xylindein from *Chlorociboria aeruginosa*	77%	23%	0%	100%

3.7. FIB

FIB images were obtained for the three pigments. They showed that all of the three extracted pigments have a surface attachment to the wood and they do not penetrate the wood cell wall. The sample of *S. cuboideum* and *C. aeruginosa* showed a similar attachment to the surface of wood, consisting of irregular void spaces between attachment points. For *S. cuboideium,* the void spaces were smaller than the ones seen for *C. aeruginosa* (see Figures 11 and 12).

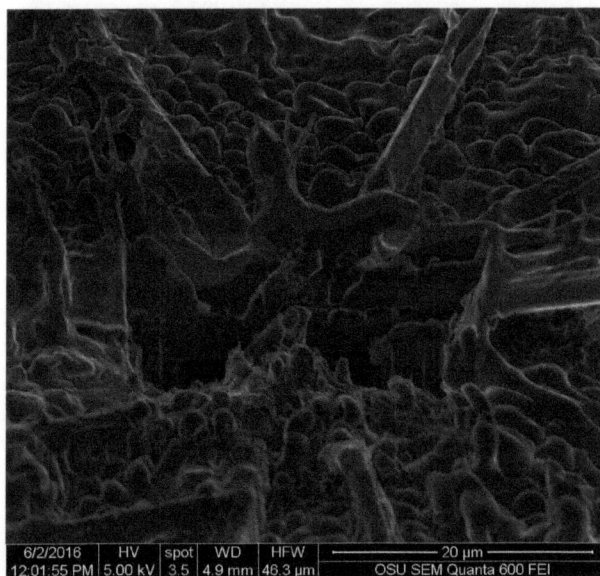

Figure 11. Cross-section of a crystal-like structure produced by *S. cuboideum* on cottonwood, showing void spaces between the attachment area of the pigment with the wood cell wall. Picture taken with an FEI QUANTA 3D dual beam SEM/FIB.

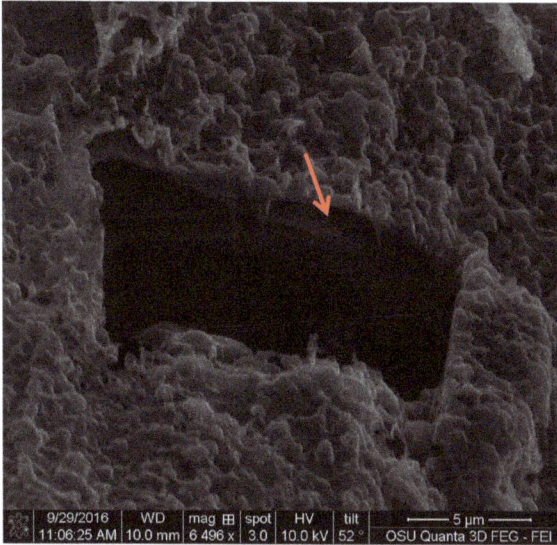

Figure 12. FIB cross-section of the extracted pigment of *C. aeruginosa* with cottonwood, showing large void spaces between the wood cell wall and the pigment, pointed to with an arrow. FEI QUANTA 3D dual beam SEM/FIB.

For *S. ganodermophthorum*, the attachment to wood was in two primary areas. In the first area it is possible to observe a continuous layer of pigment on top of the cell wall, and on the second area the surface showed a sponge-like texture (see Figure 13).

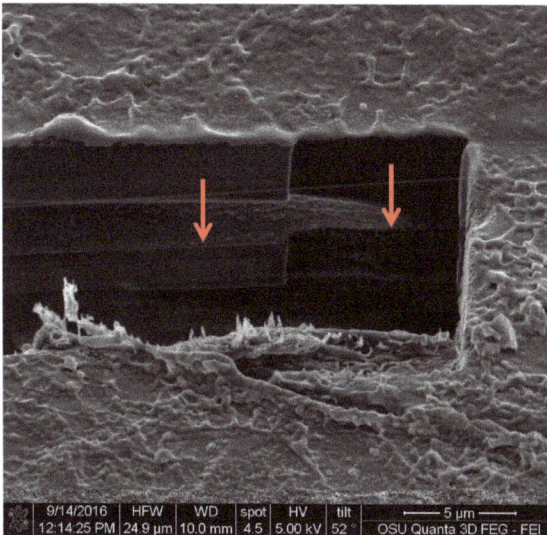

Figure 13. Cross-section of the area between the extracted pigment of *S. ganodermophthorum* and cottonwood. The left arrow shows an even attachment layer while the right arrow shows a sponge-like surface between the pigment and the wood. FEI QUANTA 3D dual beam SEM/FIB.

4. Discussion

The main distribution in wood of the extracted pigments was found in the vessels. The location of the extracted pigments contrast with naturally-secreted pigments. Naturally-secreted pigments will almost always be found in higher concentrations in wood ray cells, where the simple sugars are located [23]. Fungi develop and deposit their pigments in the ray cells of wood, as seen in the study by Tudor (melanin-producing fungi) [24] and Robinson [25] (pigmenting fungi, such as *C. aeruginosa*). Compared to other decay fungi, spalting fungi, specifically those that secrete extracellular pigments not classified as melanin, are Ascomycetes that prefer to first digest the available simple sugars in wood rays [23]. In contrast, extracted pigments, when applied by a dripping test, allocate in the vessels. This location is similar for wood coatings that are applied on the cross-section of wood [26], and likely has to do with the solutions moving through a path of least resistance.

The change of location of the pigments, from rays to vessels, could affect some of the inherent properties of the spalting fungal pigments, such as UV light resistance or color fastness. Were this the case, it could be due to differences in cellular structure. Wood ray cells can compartmentalize the pigments and hold higher concentrations of them due to their septa, in contrast to the wider and continuous vessel cells [27]. Additionally, the location of the rays in the wood can make the pigments less exposed to the effects of light, as they are organized in grouped structures. This is in contrast to the alignment of vessel elements, which, depending on the wood species, can be disperse in irregular patterns. However, recent studies on extracted fungal pigments focused on textiles [9,13] and UV light resistance [14], have shown that these pigments still retain the same properties as the ones naturally deposited by fungi.

Unfortunately the confocal microscopy method was ineffective with the samples due to the autofluorescence of wood (due to lignin and extractives) [28–31], and also because the pigments showed a similar fluorescence range. Usually, markers, like calcofluor white, are used in confocal microscopy to highlight the structures of fungal cells [32]. In the case of the extracted fungal pigments, there are no fungal cell walls where the marker can attach. This made the use of confocal microscopy not viable for further analysis of extracted fungal pigments.

Images obtained with SEM gave a detailed view of the extracted fungal pigments' topography. A study by Vega et al. on bamboo [12], showed that the pigments of *S. cuboideum* formed a rough surface on the hyphae cell wall, while microscopy on the genus *Chlorociboria* did not find a specific structure of the pigments within the hyphae and the colored areas [18]. The lack of a specific morphology was confirmed for the extracted pigments of *S. ganodermophthorum* and *C. aeruginosa*, but for *S. cuboideum*, crystal-like structures were observed. It was originally suspected that these crystals were calcium oxalates, which are commonly formed by fungi [33–35], but the EDS analysis showed no presence of calcium (Ca) in the samples. Other fungal pigments are also known to crystalize as shiraiachrome B [36], a red crystal with a perylenequinone nucleus, which differs from the described naphthoquinone composition of the red pigment draconin red from *S. cuboideum*, previously described by Golinsky et al., that only includes carbon, oxygen, and hydrogen in its structure [37]. Future studies could determine if the crystal-like structures identified on *S. cuboideum* pigments are organic single crystals or organic poly-crystals (aggrupation of single crystals) by determining if it has a diffraction pattern. This could be performed with a Kikuchi diffraction pattern analysis with a transmission electron microscope (TEM), which is broadly used for determining the crystal nature of materials [38] or with the accurate X-ray diffraction analysis [39]. If the aforementioned pigment is confirmed as a single crystal or poly-crystal, the synthesis of the pure pigment could be easier regardless of the crystal type. As a crystal compound, it would be possible to obtain a powder-like state pigment, which could allow a wider use of the pigment in diverse materials without the limitation of the solvent.

Specifically for xylindein (from *C. aeruginosa*), a compound known since the late 1800s [16,17,40–42], two clear areas were identified in the SEM images. These two areas correspond to a difference in the way that the material reflects electrons. A study by Saikawa et al. determined that xylindein produces two isomers, one that reflects the blue spectrum of light, while the second isomer reflects the yellow spectrum of light [40]. This characteristic in the xylindein molecule can explain the presence of the two areas where the electrons reflection differs. The images obtained from *S. ganodermophthorum* also presented a similar two-electron reflection area as *C. aeruginosa*, but the chemical identification of the yellow pigment is still not completed [9,15]. By the images obtained, it is possible to suspect that there might be two different molecules in the pigment, or possibly isomers.

For the textiles, *S. cuboideum* on polyester showed that the crystal-like structures wrap around the fibers of the material. This might confer better color stability on polyester than cotton by creating a more stable pigment deposition, which could make the pigments more resistant to physical and chemical abrasions. This was confirmed by Hinsch et al., who found that the red pigment performed better on polyester, which has a chemical structure similar to lignin, in terms of color fastness compared to cellulosic fibers, such as cotton [9]. Is possible to theorize that the pigments have a higher chemical affinity to lignin-like materials over the cellulose-like ones, mainly due to a higher amount of bonding sites on lignin-like compounds compared with the cellulose-like ones [27].

For the FIB analysis, the three pigments showed a similar attachment to wood. The pigments presented irregular attachment and void areas that looked similar to how adhesives attach to wood. It is possible that the pigments form similar hydrogen bonds to wood [27], and also mechanical interactions. Future research in this area should be able to confirm this using a nuclear magnetic resonance spectroscopy (NMR) or a proton nuclear magnetic resonance (HNMR) analysis.

5. Conclusions

The results obtained shed new light on how these extracted pigments deposit onto wood and textiles, and offer a view of how coating and dye manufacturers can visualize the possibilities of these fungal pigments. Extracted fungal pigments from *S. cuboideum*, *C. aeruginosa*, and *S. ganodermophthorum*, when applied on the cross-section of wood, are transported through the vessels of wood and have a surface attachment that has void and contact spaces. This may mean that there can be a chemical and mechanical interaction between the pigments and the materials. On textiles it was observed that the pigment of *S. cuboideum* wrapped around the polyester fiber, while the pigment behaved similarly on cotton as it did with wood. For base morphology, the red pigment from *S. cuboideum* forms a crystal-like structure. To confirm this, diffraction studies are needed. If a crystalline structure is confirmed, a broader use of fungal pigments can be conceived as the pigments could be turned into a powdered form that can then be applied to a wider variety of materials.

Additionally, this study determined the best methods to analyze the pigments microscopically. Scanning electron microscopy was the best method to visualize and identify the extracted fungal pigments on different materials. Light microscopy was useful to identify the allocation of the pigments in wood. Confocal microscopy was not effective to identify the attachment of pigments on wood, due to the autofluorescence of the material and the broad-range fluorescence of the pigments.

Acknowledgments: Oregon State University Electron Microscopy facilities Joan Hudson, Peter Eschbach, Teresa Sawyer, Oregon State University Center of Genome Research and Biocomputing Anne-Marie Girard Pohjanpelto. The authors gratefully acknowledge the financial support provided by the Walmart Manufacturing Innovation Fund for this work.

Author Contributions: Sarath M. Vega Gutierrez and Sara C. Robinson conceived and designed the experiments; Sarath M. Vega Gutierrez performed the experiments, analyzed the data, and wrote the paper.

Conflicts of Interest: The authors declare no conflict of interest.

References

1. Pointing, S. Feasibility of bioremediation by white-rot fungi. *Appl. Microbiol. Biotechnol.* **2001**, *57*, 20–33. [PubMed]
2. Preuss, H.P. *Pigments in Paint*; Noyes Data Corp.: Park Ridge, NJ, USA, 1974.
3. Carrubba, R.V.; Golden, R.L. Preparation of Anilines from Oxygenated Cyclohexanes. Google Patents US3347921A, 17 October 1967.

4.   Oda, H. An attempt to improve the light fastness of natural dyes. Effect of UV absorbers on the photofading of red carthamin. *Jpn Home Econ. J.* **2002**, *53*, 271–277.

5.   Erdoğrul, Ö.; Azirak, S. Review of the studies on the red yeast rice (*Monascus purpureus*). *Turk. Electron. J. Biotechnol.* **2004**, *2*, 37–49.

6.   Ahmad, W.A.; Ahmad, W.Y.W.; Zakaria, Z.A.; Yusof, N.Z. Application of bacterial pigments as colorant. In *Application of Bacterial Pigments as Colorant*; Springer: Berlin, Germany, 2012; pp. 57–74.

7.   Ferreira, E.S.; Hulme, A.N.; McNab, H.; Quye, A. The natural constituents of historical textile dyes. *Chem. Soc. Rev.* **2004**, *33*, 329–336. [CrossRef] [PubMed]

8.   Hinsch, E.M.; Weber, G.; Chen, H.-L.; Robinson, S.C. Colorfastness of extracted wood-staining fungal pigments on fabrics: A new potential for textile dyes. *J. Text. Appar. Technol. Manag.* **2015**, *9*, 3.

9.   Cristea, D.; Vilarem, G. Improving light fastness of natural dyes on cotton yarn. *Dyes Pigments* **2006**, *70*, 238–245. [CrossRef]

10.   Hinsch, E.M. A Comparative Analysis of Extracted Fungal Pigments and Commercially Available Dyes for Colorizing Textiles. Mater's Thesis, Oregon State University, Corvallis, OR, USA, 2015.

11.   Robinson, S.C.; Weber, G.; Hinsch, E.; Vega Gutierrez, S.M.; Pittis, L.; Freitas, S. Utilizing extracted fungal pigments for wood spalting: A comparison of induced fungal pigmentation to fungal dyeing. *J. Coat.* **2014**, *2014*, 1–8. [CrossRef]

12.   Vega Gutierrez, S.; Vega Gutierrez, P.; Godinez, A.; Pittis, L.; Huber, M.; Stanton, S.; Robinson, S. Feasibility of coloring bamboo with the application of natural and extracted fungal pigments. *Coatings* **2016**, *6*, 37. [CrossRef]

13.   Weber, G.; Chen, H.L.; Hinsch, E.; Freitas, S.; Robinson, S. Pigments extracted from the wood-staining fungi *Chlorociboria aeruginosa*, *Scytalidium cuboideum*, and *S. ganodermophthorum* show potential for use as textile dyes. *Color. Technol.* **2014**, *130*, 445–452. [CrossRef]

14.   Beck, H.G.; Freitas, S.; Weber, G.; Robinson, S.C.; Morrell, J.J. Resistance of fungal derived pigments to ultraviolet light exposure. In *International Research Group in Wood Protection*; IRG/WP: St George, UT, USA, 2014.

15.   Weber, G.; Boonloed, A.; Naas, K.M.; Koesdjojo, M.T.; Remcho, V.T.; Robinson, S.C. A method to stimulate production of extracellular pigments from wood-degrading fungi using a water carrier. *Curr. Res. Environ. Appl. Mycol.* **2016**, *6*, 218–230.

16.   Edwards, R.L.; Kale, N. The structure of xylindein. *Tetrahedron* **1965**, *21*, 2095–2107. [CrossRef]

17.   Blackburn, G.M.; Ekong, D.E.; Nielson, A.H.; Todd, L. Xylindein. *Chimia* **1965**, *19*, 208–212.

18.   Blanchette, R.A.; Wilmering, A.M.; Baumeister, M. The use of green-stained wood caused by the fungus *Chlorociboria* in intarsia masterpieces from the 15th century. *Holzforschung* **1992**, *46*, 225–232. [CrossRef]

19.   Michaelsen, H.; Unger, A.; Fischer, C.H. Blaugrüne färbung an intarsienhölzern des 16. Bis 18. Jahrhunderts. Available online: https://opus4.kobv.de/opus4-fhpotsdam/frontdoor/index/index/docId/1247 (accessed on 10 November 2016).

20.   Robinson, S.C.; Hinsch, E.; Weber, G.; Freitas, S. Method of extraction and resolubilisation of pigments from *Chlorociboria aeruginosa* and *Scytalidium cuboideum*, two prolific spalting fungi. *Color. Technol.* **2014**, *130*, 221–225. [CrossRef]

21.   Maeda, M.; Yamauchi, T.; Oshima, K.; Shimomura, M.; Miyauchi, S.; Mukae, K.; Sakaki, T.; Shibata, M.; Wakamatsu, K. Extraction of xylindein from *Chlorociboria aeruginosa* complex and its biological characteristics. *Bull. Nagaoka Univ. Technol.* **2003**, *25*, 105–111.

22.   Robinson, S.C.; Hinsch, E.; Weber, G.; Leipus, K.; Cerney, D. Wood colorization through pressure treating: The potential of extracted colorants from spalting fungi as a replacement for woodworkers' aniline dyes. *Materials* **2014**, *7*, 5427–5437. [CrossRef]

23.   Zabel, R.A.; Morrell, J.J. *Wood Microbiology. Decay and Its Prevention*; Harcourt Brace Jovanovich, Academic Press, Inc.: New York, NY, USA, 1992.

24.   Tudor, D.; Robinson, S.C.; Sage, T.L.; Krigstin, S.; Cooper, P.A. Microscopic investigation on fungal pigment formation and its morphology in wood substrates. *Open Mycol. J.* **2014**, *8*, 174–186. [CrossRef]

25.   Robinson, S.C.; Michaelsen, H.; Robinson, J.C. *Spalted Wood. The History, Science and Art of a Unique Material*, 1st ed.; Schiffer Publishing, Ltd: Atglen, PA, USA, 2016; p. 287.

26.   De Meijer, M.; Thurich, K.; Militz, H. Comparative study on penetration characteristics of modern wood coatings. *Wood Sci. Technol.* **1998**, *32*, 347–365. [CrossRef]

27. Shmulsky, R.; Jones, D. *Forest Products and Wood Science*, 6th ed.; John Wiley & Sons: Hoboken, NJ, USA, 2011.

28. Billinton, N.; Knight, A.W. Seeing the wood through the trees: A review of techniques for distinguishing green fluorescent protein from endogenous autofluorescence. *Anal. Biochem.* **2001**, *291*, 175–197. [CrossRef] [PubMed]

29. Olmstead, J.A.; Gray, D.G. Fluorescence emission from mechanical pulp sheets. *J. Photochem. Photobiol. A Chem.* **1993**, *73*, 59–65. [CrossRef]

30. Olmstead, J.; Gray, D. Fluorescence spectroscopy of cellulose, lignin and mechanical pulps: A review. *J. Pulp Pap. Sci.* **1997**, *23*, J571–J581.

31. Albinsson, B.; Li, S.; Lundquist, K.; Stomberg, R. The origin of lignin fluorescence. *J. Mol. Struct.* **1999**, *508*, 19–27. [CrossRef]

32. Hickey, P.C.; Swift, S.R.; Roca, M.G.; Read, N.D. Live-cell imaging of filamentous fungi using vital fluorescent dyes and confocal microscopy. *Methods Microbiol.* **2004**, *34*, 63–87.

33. Whitney, K.D.; Arnott, H.J. Calcium oxalate crystal morphology and development in agaricus bisporus. *Mycologia* **1987**, 180–187. [CrossRef]

34. Dutton, M.V.; Evans, C.S.; Atkey, P.T.; Wood, D.A. Oxalate production by basidimycetes, including the white rot species *Coriolus versicolor* and *Phanerochaete chrysosporium*. *Appl. Microbiol. Biotechnol.* **1993**, *39*, 5–10. [CrossRef]

35. Dutton, M.V.; Kathiara, M.; Gallagher, I.M.; Evans, C.S. Purification and characterisation of oxalate decarboxylase from *Coriolus versicolor*. *FEMS Microbiol. Lett.* **1994**, *116*, 321–325. [CrossRef]

36. Wu, H.; Lao, X.F.; Wang, Q.W.; Lu, R.R.; Shen, C.; Zhang, F.; Liu, M.; Jia, L. The shiraiachromes: Novel fungal perylenequinone pigments from shiraia bambusicola. *J. Nat. Prod.* **1989**, *52*, 948–951. [CrossRef]

37. Golinski, P.; Krick, T.P.; Blanchette, R.A.; Mirocha, C.J. Chemical characterization of a red pigment (5,8-dihydroxy-2,7-dimethoxy-1,4-naphthalenedione) produced by arthrographis cuboidea in pink stained wood. *Holzforschung* **1995**, *49*, 407–410. [CrossRef]

38. Morito, S.; Tanaka, H.; Konishi, R.; Furuhara, T.; Maki, T. The morphology and crystallography of lath martensite in Fe-C alloys. *Acta Mater.* **2003**, *51*, 1789–1799. [CrossRef]

39. Woolfson, M. *X-ray Crystallography*; Cambridge University Press: Cambridge, UK, 1970.

40. Saikawa, Y.; Watanabe, T.; Hashimoto, K.; Nakata, A. Absolute configuration and tautomeric structure of xylindein, a blue-green pigment of *Chlorociboria* species. *Phytochemistry* **2000**, *55*, 237–240. [CrossRef]

41. Gümbel, W. Über xylindein. *Berichte der Deutschen Chemischen Gesellschaft* **1858**, *41*, 113–115.

42. Lieberman, C. Über xylindein. *Berichte der Deutschen Chemischen Gesellschaft* **1874**, *7*, 1102–1103.

Journal of
Fungi

MDPI

Article

Perstraction of Intracellular Pigments through Submerged Fermentation of *Talaromyces* spp. in a Surfactant Rich Media: A Novel Approach for Enhanced Pigment Recovery

Lourdes Morales-Oyervides [1,2], Jorge Oliveira [1], Maria Sousa-Gallagher [1], Alejandro Méndez-Zavala [2] and Julio Cesar Montañez [2,*]

[1] School of Engineering, University College Cork, Cork, Ireland; lourdesmorales@uadec.edu.mx (L.M.-O.); j.oliveira@ucc.ie (J.O.); M.deSousaGallagher@ucc.ie (M.S.-G.)
[2] Department of Chemical Engineering, Universidad Autónoma de Coahuila, Saltillo 25280, Mexico; alejandro.mendez@uadec.edu.mx
* Correspondence: julio.montanez@uadec.edu.mx; Tel.: +52-844-416-9213

Received: 1 June 2017; Accepted: 22 June 2017; Published: 27 June 2017

Abstract: A high percentage of the pigments produced by *Talaromyces* spp. remains inside the cell, which could lead to a high product concentration inhibition. To overcome this issue an extractive fermentation process, perstraction, was suggested, which involves the extraction of the intracellular products out of the cell by using a two-phase system during the fermentation. The present work studied the effect of various surfactants on secretion of intracellular pigments produced by *Talaromyces* spp. in submerged fermentation. Surfactants used were: non-ionic surfactants (Tween 80, Span 20 and Triton X-100) and a polyethylene glycerol polymer 8000, at different concentrations (5, 20, 35 g/L). The highest extracellular pigment yield (16 OD_{500nm}) was reached using Triton X-100 (35 g/L), which was 44% higher than the control (no surfactant added). The effect of addition time of the selected surfactant was further studied. The highest extracellular pigment concentration (22 OD_{500nm}) was achieved when the surfactant was added at 120 h of fermentation. Kinetics of extracellular and intracellular pigments were examined. Total pigment at the end of the fermentation using Triton X-100 was 27.7% higher than the control, confirming that the use of surfactants partially alleviated the product inhibition during the pigment production culture.

Keywords: fungal pigments; perstraction; surfactant; *Talaromyces*

---

1. Introduction

In recent years, consumer trends have been oriented to natural products, especially in the food industry. Consumer expectations have led the food processing industry to improve processes in order to deliver high quality-products [1]. To meet the requirements for safer food and also satisfy consumer preferences, scientists are searching for new food additives and new food processing methods [2]. The application of food additives will depend on their function in foods and can be classified as preservatives, nutritional additives, flavouring agents, texturising agents and colouring agents [3]. Natural colourants are obtained from sources like plants [4,5], insects [6] and microorganisms [7]. However, application of microbial pigments still represents a major challenge to biotechnology due to lower extraction or production yields. Dufosse et al. (2014) [8] emphasised in their fungal pigments review, the crucial role that filamentous fungi are currently playing as microbial cell factories, mainly due to the attractive range of stable bio-colourants that they are able to synthesise under controlled conditions. The most-well documented pigment producer fungi is *Monascus* [9]. However, in the last years, pigment producing fungi from other species have gained more attention due to the mycotoxin (citrinin) produced by *Monascus* [10].

Fungi belonging to the genus *Talaromyces* (formerly *Penicillium*) have attracted the attention of scientists due to their high pigment production yields [11,12] and moreover, due to the high thermal stability [13], antioxidant properties [14], antibacterial properties [15] and the absence of toxicity of these pigments [15,16].

However, their successful industrial application will not only depend on their safety or added value properties. The economics of the process need to be assessed early in the design stage before going through more expensive research, such as the scale-up of the process or the characterisation of the molecules. There are various strategies to achieve a cost-effective process such as using cheaper substrates to reduce raw material costs, optimising process conditions to achieve high yields or implementing an efficient product recovery. From these, developing an efficient downstream processing seems to be a rapid strategy to maximise product recovery and to reduce costs.

Pigments produced by *Talaromyces* are excreted out of the cell. However, most remain inside the cell, adding several unit operations to the downstream processing such as cell disruption, extraction of intracellular products, followed by the cell debris removal. Furthermore, a high intracellular pigment concentration could lead to product concentration inhibition. To overcome this issue researchers have been applying an extractive fermentation process, "perstraction", where the intracellular products are transported out of the cell by using a two-phase system during the fermentation, where the fermentation media represents one of the phases and the other should be an extractive solution. This process has been used in microalgae cultures [17,18] and is also known as "milking". Ziolkowska and Simon, (2014) [17] reported that "milking algae" is a new successful technology that allows for the reduction of process costs by the extraction of intracellular products, increasing yields and reduction of unit operations. Two aspects are of significant importance for the selection of a suitable extractant for the perstraction of an intracellular product [19]: the biocompatibility of the solvent with the microorganism allowing cell growth, and that the extractant should induce permeability of the cell wall. Recent studies have reported the application of non-ionic surfactant as an effective extractant on the perstraction of intracellular pigments produced by the fungus *Monascus* [20,21]. Wang et al. (2013) [22] reported that the use of non-ionic surfactant Triton X-100 highly increased the pigment production and the extracellular/intracellular pigment ratio by *Monascus* in submerged fermentation. A specific characteristic of the non-ionic aqueous solution is the formation of the cloud point system. The addition of non-ionic surfactant to an aqueous solution forms micelles and when this solution is above the surfactant cloud point, the micelle aqueous solution separates into two phases, a surfactant rich phase (coacervate phase) and a dilute phase [23]. The application of the cloud point system has been reviewed as an advantageous pre-concentration step prior to purification in many bioprocesses, such as extraction of fatty acids from microalgae cultures [18]; extractive fermentation of proteins produced by bacteria [24]; production of L-phenylacetylcarbinol by *Sacharomyces cerevisiae* [25] and in exporting intracellular pigments produced by *Monascus* into its fermentation broth [20,22,26].

The application of this concept to submerged fermentation of *Talaromyces* spp. has not been studied, and therefore, the objective of this work was to analyse the effect of various surfactants, at different concentrations, on secreting the intracellular pigments produced by *Talaromyces* spp. into the fermentation broth. The relevance of the precise time of addition of the surfactant was further studied in order to define conditions for maximum extracellular pigments recovery. In addition, the kinetics of pigment production, biomass and substrate consumption during *Talaromyces* spp. culture using non-ionic surfactant micelle aqueous solution were analysed.

2. Materials and Methods

2.1. Microorganism

Talaromyces spp. was used for the production of red pigments (DIA-UAdeC). The purified strain had been previously isolated and characterised as *Penicillium purpurogenum* GH2 [27,28]. *Penicillium purpurogenum* has, however, been transferred to *Talaromyces* spp. [29]. The strain was maintained on PDA (Potato dextrose agar) slants at 4 °C and sub-cultured periodically.

2.2. Culture Media

The PDA medium was prepared with a concentration of 39.0 g/L (Bioxon, Mexico). The medium Potato Dextrose Broth (PDB medium, ATCC medium: 336) was prepared by boiling 0.3 kg of finely diced potatoes in 500 mL of water until thoroughly cooked; then the potatoes were filtered through cheesecloth and water was added to the filtrate to complete a volume of 1.0 L. Finally, 20.0 g of glucose is added before sterilisation. The Czapek-dox modified medium [30] consisted in (g/L): D-xylose 15.0, NaNO$_3$ 3.0, MgSO$_4$·7H$_2$O 0.5, FeSO$_4$·7H$_2$O 0.1, K$_2$HPO$_4$ 1.0, KCl 1.0 and ethanol 20.0.

2.3. Inoculum Preparation

PDB medium was used for the inoculum preparation. Flasks (125 mL) containing 25 mL of PDB medium were sterilised and then inoculated with a spore suspension (1 × 10^5 spores/mL) of *Talaromyces* spp. previously incubated for 5 days. The flasks were then incubated at 30 °C for 84 h in an orbital shaker (Innova 94, New Brunswick Scientific, Edison, NJ, USA) at 200 rpm [11,31].

2.4. Cultivation Conditions

Czapek-dox medium was used for pigment production. The initial pH of the Czapek-dox modified medium was adjusted to 5 before sterilising by using 0.22 m sterile membranes (Millipore, Billerica, MA, USA).

Four different surfactants were tested, assessing their effect on the pigment production and growth in submerged fermentation of *Talaromyces* spp. The surfactants analysed were Tween 80, Span 20, Triton X-100 and polyethylene glycerol polymer PEG 8000, at 3 different concentrations (5, 20, 35 g/L). Surfactants were added at the beginning of fermentation. Extractive fermentation was carried out in 125 mL Erlenmeyer flasks containing 25 mL of medium. Flasks were inoculated with a mycelial suspension 10% (v/v). The inoculated flasks were incubated at 30 °C in an orbital shaker (Innova 94, New Brunswick Scientific, Edison, NJ, USA) at 200 rpm for 8 days. Control experiments were also performed with no surfactant added.

2.5. Analytical Methods

Extracellular pigment recovery was performed according to the methodology reported by Méndez-Zavala et al. (2011) [32]. Each sample was centrifuged at 8000 rpm for 20 min at 4 °C (Sorvall, Primo R Biofuge Centrifugation Thermo, Waltham, MA, USA). The supernatant was then filtered through a 0.45 µm cellulose filter (Millipore, USA). Recovered mycelia were rinsed with distilled water until the supernatant was cleared and used for further extraction of intracellular pigments [20]. Mycelia were soaked in 25 mL of a 70% (v/v) ethanol aqueous solution which represented the initial fermentation broth volume. Extraction was carried out on an orbital shaker (Inova 94, New Brunswick Scientific, USA) at 200 rpm at 30 °C for 1 h. The concentration of red pigments (extracellular and intracellular) was quantified indirectly by simply measuring the optical density at 500 nm using a spectrophotometer (Cary 50, UV-Visible Varian, Palo Alto, CA, USA). The biomass concentration was determined using the gravimetric method. The analysis of substrate consumption was determined by quantifying the total sugar content using the method reported by [33]. The experiments were replicated three times.

2.6. Cloud Point Extraction

Extracellular pigment was separated into a two-phase cloud point system: a surfactant-rich phase and a dilute phase [18]. Pigments-surfactant solution was heated until the formation of these two phases (70 °C) and until reaching the equilibrium (30 min).

An estimation of the pigments that remained in each phase (diluted and coacervate) was done by performing a spectral analysis within the visible wave range (Cary 50, UV-Visible Varian). Various authors have expressed fungal pigments concentration as optical density units at the absorption maxima for a specific wavelength: 400–420 (yellow), 450–470 (orange) and red (490–510) [12,20,26,34,35].

2.7. Kinetic Parameters

Pigment extraction productivity ($P_{Y/t}$, OD/h), biomass productivity ($P_{B/t}$, g/L/h) and yield of product per unit of biomass ($Y_{Y/B}$, OD·L/h) were determined by the following equations, [36]:

$$P_{\frac{Y}{t}} = \frac{Y_\infty - Y_0}{t_\infty - t_0} \tag{1}$$

$$P_{\frac{B}{t}} = \frac{B_\infty - B_0}{t_\infty - t_0} \tag{2}$$

$$Y_{\frac{Y}{B}} = \frac{Y_\infty - Y_0}{B_\infty - B_0} \tag{3}$$

where Y_0 and B_0 correspond to the yield of intracellular pigment (OD_{500nm}) and biomass (g/L) at the time (t_0, min) when the surfactant (Triton X-100) was added; Y_∞ and B_∞ are the yield of pigments (OD_{500nm}) and biomass (g/L) at the end of the fermentation (t_∞, min).

2.8. Data Analysis

Kinetic parameters were set from experimental data. Statistical analyses were made with Statistica 7.0 software (StatSoft, Tulsa, OK, USA).

3. Results and Discussion

3.1. Effect of Surfactants on Secretion of Intracellular Pigments into Fermentation Broth (Screening)

As a first step towards developing a non-ionic surfactant micelle aqueous solution as a fermentation media, a screening of the non-ionic surfactants most commonly used in bioprocess was performed [22,23]. Selection of a suitable surfactant was based on the highest pigment production and extracellular/intracellular pigment ratio. Figure 1 shows the extracellular and intracellular pigments obtained after eight days of fermentation for each surfactant, evaluated at three different concentrations (5, 20 and 35 g/L).

Results for the control media indicated that a high extracellular pigment yield was reached and nearly 40% of the total pigment was still remaining inside of the cell. *Talaromyces* spp. was capable of growing in all surfactants studied; however, pigment production was strongly affected by the addition of surfactant at the beginning of the fermentation. The addition of PEG 8000 caused a reduction in pigment production by more than 50%, in comparison with the control media. This finding limits the use of PEG as a water soluble polymer in aqueous two-phase systems (ATPS) on in-situ product removal processes [37]. When Span 20 was added, the total pigment level was reduced by 80%, in comparison with control media. It was also observed that mycelia morphology changed (dispersed mycelia) and led to a higher biomass growth. Similar behaviour (higher cell growth) was reported [38] on β-Carotene production by *Blakeslea trispora* (Mating type, + and −) with the addition of Span 20. Biomass increment may be due to the change in morphology, i.e., fragmented mycelium led to a major aeration, conditions that allowed higher growth, but were not optimal for pigment production. The addition of Tween 80 at low concentration (5 g/L) caused a reduction of only less than 10% of the total pigment, however, the extracellular/intracellular pigment ratio was increased from 1.72 (control) to 1.79. This may be an indication of the permeability induced on the cell wall by Tween 80; however, increasing the levels of this surfactant resulted in a reduction in both biomass and total pigment, showing the low biocompatibility of this surfactant with *Talaromyces* spp. Similarly, Zhang et al. (2013) [39] reported that the addition of a high concentration of Tween 80 during the production of Antrodin C in submerged fermentation by *A. camphorata* led to massive damage on the cell membrane. When Triton X-100 surfactant was added (35 g/L), an enhancement of extracellular pigment production level (44%) was observed. Total pigment production (extracellular and intracellular) was not statistically different from the control experiments (without addition of surfactant); however, extracellular/intracellular pigment ratio was increased from 1.73 (control) to 11.75. Biomass reduction with elevated concentration of Triton X-100 may indicate that this non-ionic surfactant has a low biocompatibility with the strain, although it may also be related to the secretion of the intracellular pigments. A cell with an elevated concentration of pigment presents a higher weight than an "empty cell". Triton X-100 has been successfully applied on the secretion of intracellular pigments of *Monascus* [20,26,40], similarly, Triton X-100 has presented high biocompatibility with microbial cell allowing modification of membrane to induce permeability of certain desired metabolites [21,41]. Based on the above results, Triton X-100 (3.5%) was selected for subsequent studies.

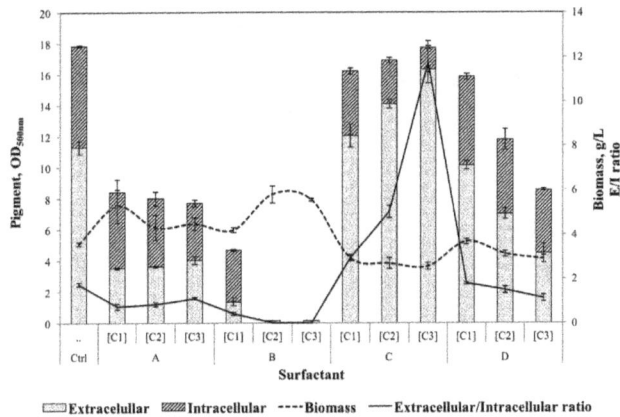

Figure 1. Screening of surfactants' effect on extracellular and intracellular pigment yield, extracellular/intracellular pigment ratio and growth. Ctrl) Without surfactant, (**A**) polyethylene glycerol polymer PEG 8000; (**B**) Span 20; (**C**) Triton X-100 and (**D**) Tween 80. [Ci, g/L]: C1, 5; C2, 20; C3, 35.

3.2. Effect of Addition Time of Triton X-100 during the Fermentation

In the previous experiments, surfactants were added at the beginning of the fermentation. Many authors have added surfactants at different stages of the fermentation to enhance microbial production [39,41]; however, the best time for surfactant addition will vary depending on the microorganism. In order to fully comprehend the effect of Triton X-100 on biomass growth and to enhance pigments production by *Talaromyces* spp., 35 g/L was added at different fermentation times (0, 24, 48, 72, 96, 120 and 144 h). Results for final biomass, extracellular and intracellular pigments are presented in Figure 2.

Biomass was reduced by 20% when Triton X-100 was added in the interval of 0–3 days and decayed to 50% when it was added at later stages of the fermentation, in comparison with the control. Wang et al. (2013) [22] reported that the addition of Triton X-100 to pigments production by *Monascus* at later stages of the fermentation process resulted in higher biomass growth. In contrast, in this study a consequent addition time of Triton X-100 did not represent better conditions to biomass growth; thus, the results obtained were not sufficient to demonstrate a noticeable Triton X-100 inhibitory effect on the cell. A time course study for each addition time could give more understanding of the Triton X-10 harmful effect on the cell growth, i.e., extended lag phase, reduced maximum biomass and reduced growth rate. However, the main objective of this study was to enhance pigment production and to increase extracellular/intracellular pigment ratio. Figure 2 shows that the excretion of intracellular pigments was considerably enhanced with the addition of Triton X-100; extracellular pigment increased gradually with the time of addition reaching the maximum extracellular pigment (22.1 ± 1.2 OD_{500nm}) when Triton X-100 was added at 120 h of fermentation, which corresponds to an increment of 90%, as compared to the control. Furthermore, extracellular/intracellular pigment ratio increased from 2.5 ± 0.1 to 37.9 ± 5.2 for control and Triton X-100 addition at 120 h of fermentation process, respectively. These results are in agreement with previous studies reported on *Monascus pigments* [20,26,40], where it was determined that adding Triton X-100 at a late stage of the fermentation process stimulated intracellular pigment excretion. Based on the above results, the addition of Triton X-100 at 120 h of fermentation at a concentration of 35 g/L was selected for further studies.

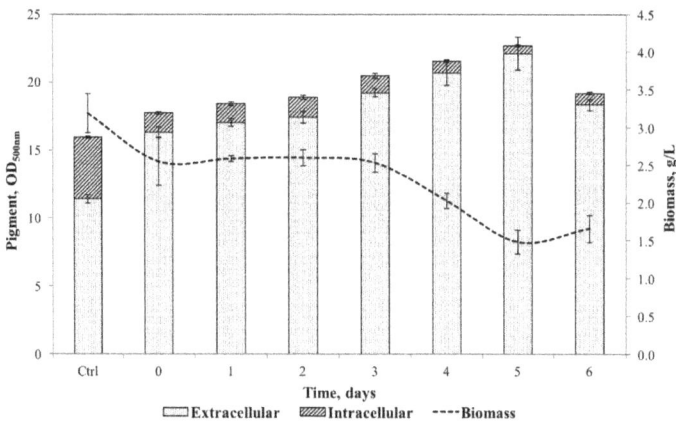

Figure 2. Effect of the addition time of Triton X-100 (35 g/L) on extracellular and intracellular pigment yield and biomass growth.

3.3. Pigments Production Kinetics of Talaromyces spp. by Using Triton X-100

Kinetics of biomass growth, pigment production and substrate consumption with and without the addition of Triton X-100 at 120 h of fermentation are depicted in Figure 3 (a and b respectively). Figure 3a shows that there is no lag phase for biomass growth; the microorganism started to grow (exponential phase) from 0 to 120 h, then reached a short stationary phase (120 to 150 h), after which biomass weight decayed. There was a slight relation between growth and pigment production.

Production of pigments started at 24 h and increased gradually until the end of fermentation. It is interesting to note that there is a correlation between extracellular, intracellular pigments and biomass decay. Extracellular and intracellular pigment profiles incremented in a similar pattern until 120 h. After 120 h of submerged fermentation, intracellular pigments decreased correspondingly with the biomass decay and further increment of the extracellular pigments. This behaviour led us to hypothesise that when the fungus reaches the stationary phase (120 h), it stops producing pigments and the further increase of extracellular pigments is partially due to the excretion of intracellular pigments out the cell. This excretion of intracellular pigments may contribute to the biomass decay after 120 of fermentation.

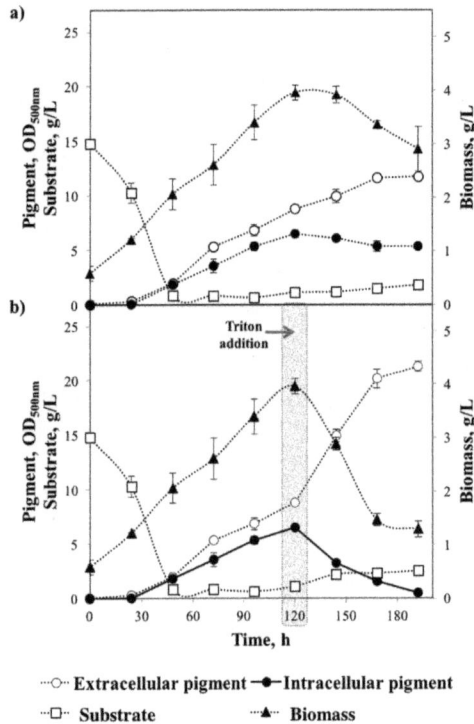

Figure 3. Kinetics of extracellular and intracellular pigment production, substrate consumption and biomass growth. (**a**) Without surfactant; (**b**) Triton X-100 added at 120 h of fermentation.

At 120 h of fermentation process, the fungus had reached the stationary phase and thus the addition of Triton X-100 did not cause biomass growth inhibition; this was the reason for the highest extracellular yields being obtained when Triton X-100 was added at this stage. The above results suggested that the surfactant only induced cell permeability and allowed intracellular pigments excretion. Previous work on excretion of *Monascus pigments* have suggested that the addition of Triton X-100 increases the permeability and fluidity of the cell due to the surfactant effect on cell membrane lipids [22]. Furthermore, Deive et al. (2009) [42] reported that the addition of surfactants increased the lipids solubility on the cell membrane changing the cell mass transfer.

The major effect on the cell of the addition of Triton X-100 to the fermentation process could be reflected in the elevated increment of extracellular pigments. Extracellular pigments rose from 8.8 ± 0.2 to 14.9 ± 0.56 OD_{500nm} after 24 h of Triton X-100 addition, which showed this increment as being 5.3 times higher than the control. Furthermore, the extracellular/intracellular pigment ratio was highly increased after 24 h Triton X-100 addition, in comparison with the control, from 1.6 to 4.6. A remarkable observation is that after Triton X-100 addition (120 h), the correlation between extracellular pigments accumulation, intracellular pigments decrease and biomass decay was similar in comparison to the control. This suggests that when the fungus has reached the stationary phase, extracellular pigment increment is only due to the release of intracellular pigments contributing to the higher reduction in biomass weight observed when using Triton X-100 than with the control.

In order to correlate the extraction of intracellular pigments to biomass decay; the pigment extraction productivity ($P_{Y/t}$, OD/h) and biomass productivity ($P_{B/t}$, g/L/h) was obtained after the addition of Triton X-100 for both kinetics (control and perstraction processes). Table 1 shows the parameters obtained. $P_{Y/t}$ and $P_{B/t}$ were 400% higher when using Triton X-100 in comparison with control (negative signs indicate intracellular pigments reduction and biomass decay).

Table 1. Intracellular realising (K_I), biomass decay (K_B) rates from 120 h to 168 h of fermentation kinetics and K_I/K_B ratio. (**A**) Control, without surfactant; (**B**) Addition of Triton X-100.

Kinetics	Rate, k		Ratio
	Intracellular, OD_{500}/h	Biomass, g/L/h	K_I/K_B, OD_{500}, L/g
A	-0.023 ± 0.001	-0.0125 ± 0.002	1.90 ± 0.19
B	-0.103 ± 0.004	-0.052 ± 0.004	1.97 ± 0.10

The yield of intracellular product released per unit of biomass was calculated at 120 h of fermentation (time of Triton addition) for both kinetics. Statistical analysis showed that there were no significant differences (95% confidence level) between the $Y_{Y/B}$ obtained in the control and after the addition of Triton X-100. In both cases the cell was releasing approximately 1.90 OD_{500nm} per gram of biomass per litre.

Biomass decay (cell death) has represented a critical concept in microbiology and its estimation is continuously approached by different kinetic models [43]. Van Bodegom (2007) [44] mentioned that there are some inconsistencies related to cell death, because its quantification does not distinguish between biomass losses related to cell lysis or to the actual transport phenomena of intracellular products outside of the cell. In order to understand the mechanism of pigments release and biomass decay relationship, more studies need to be done, due to the variety of components present in the cell.

Regarding substrate consumption, there was no effect when Triton X-100 was added, as the addition occurred after substrate depletion. At the end of the fermentation process (192 h), extracellular pigments were enhanced above 80% with the addition of Triton X-100 at the end of the exponential growth phase (120 h). Visual kinetics for extracellular pigments enhancement are illustrated in Figure 4. Similar results were obtained by Wang et al. (2013) [22] who reported an increment of 88.4% of pigments produced by *Monascus* by adding Triton X-100 at a late stage of the submerged fermentation. Furthermore, total pigment (extracellular pigment plus intracellular pigment) at the end of the fermentation was 27.7% higher with Triton X-100 compared to control.

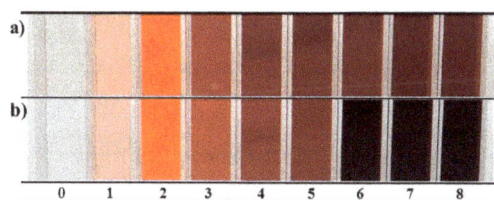

Figure 4. Visual kinetics of colour enhancement. X-axis in days. (**a**) Without surfactant; (**b**) Triton X-100 added at 120 h of fermentation.

3.4. Pigments Partitioning in Cloud Point System

The inner figure in Figure 5 shows the phase separation of the micelle aqueous solution when it reached a temperature above its cloud point (70 °C). It can be seen that the extracellular pigments obtained were separated between the dilute and coacervate phase.

Spectral analysis (Figure 5) shows a maximum absorbance at 414 and 500 nm for all the analysed samples. According to previous studies for *Talaromyces* pigments [12,34,45] these peaks correspond to yellow and red, respectively. Optical density values for each peak are also shown in Figure 5.

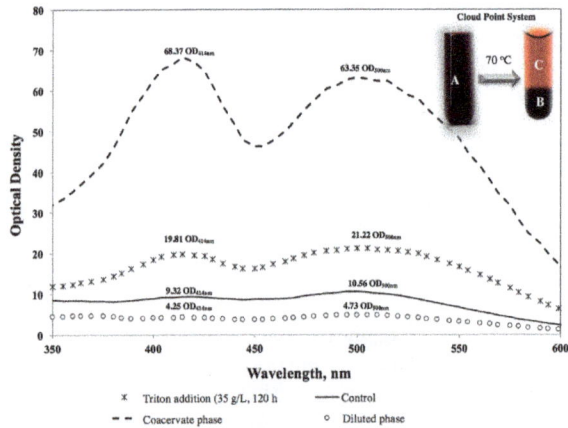

Figure 5. Spectral analysis within the visible wave range performed on the extracellular pigments obtained. Inner Figure: (**A**) the addition of Triton X-100 (35 g/L) at 120 h of fermentation, (**B**) coacervate phase, (**C**) diluted phase.

Red pigments concentration was much higher in the coacervate phase than in the diluted phase (13 times higher). Yellow pigments showed more affinity with the coacervate phase than red pigments, as its concentration was 16 times higher in the coacervate than in the diluted phase. The concentration factor, defined as the ratio of the pigment concentration in the coacervate phase to the concentration in the micelle aqueous solution [24], was 2.98 and 3.45 for red and yellow pigments, respectively. Hu et al. (2012) [20] confirmed the same affinity to the coacervate phase of pigments produced by *Monascus* in a cloud point system by Thin-layer chromatography (TLC) analysis.

4. Conclusions

In this study, applying a biphasic fermentation system was shown as a novel and effective bioprocess strategy for an in-situ extraction of intracellular pigments during fermentation, increasing the overall production. Among the studied surfactants, nonionic surfactant Triton X-100 was the most suitable for an in-situ extraction of *Talaromyces* pigments. Under the optimal perstraction conditions, the maximum extracellular pigment yield was 80% higher than the control. Total pigment enhancement (27.7%) confirmed that the addition of Triton X-100 had at least partially alleviated the product inhibition.

Maximising product recovery was evident, and it was demonstrated that the application of the cloud point system could be an advantageous pre-concentration step prior to purification of the pigments produced by *Talaromyces* spp. However, more studies are needed at large scale aired-bioreactors to assess if foaming issues must be addressed.

Also, a techno-economic analysis of the process using perstraction is required to verify if the extraction of pigments during the fermentation reduces, in fact, the costs related to the downstream processing. In any case, using a surfactant rich media can substitute the use of organic solvents for the cell disruption and extraction processes during the downstream processing.

Acknowledgments: Author L. Morales-Oyervides acknowledges CONACyT-México for the financial support provided for conducting her PhD studies (215490-2011).

Author Contributions: Julio Cesar Montañez and Jorge Oliveira conceived and designed the experiments; Lourdes Morales-Oyervides performed the experiments; Jorge Oliveira and Maria Sousa-Gallagher analysed the data; Alejandro Méndez-Zavala and Julio Cesar Montañez contributed reagents/materials/analysis tools; Lourdes Morales-Oyervides wrote the paper.

Conflicts of Interest: The authors declare no conflict of interest.

References

1. Grunert, K.G. Food quality and safety: Consumer perception and demand. *Eur. Rev. Agric. Econ.* **2005**, *32*, 369–391. [CrossRef]
2. El-Wahab, H.M.; Moram, G.S. Toxic effects of some synthetic food colorants and/or flavor additives on male rats. *Toxicol. Ind. Health* **2013**, *29*, 224–232. [CrossRef] [PubMed]
3. Carocho, M.; Barreiro, M.F.; Morales, P.; Ferreira, I.C. Adding molecules to food, pros and cons: A review on synthetic and natural food additives. *Compr. Rev. Food Sci. Food Saf.* **2014**, *13*, 377–399. [CrossRef]
4. Onslow, M.W. *The Anthocyanin Pigments of Plants*; Cambridge University Press: Cambridge, UK, 2014.
5. Boo, H.O.; Hwang, S.J.; Bae, C.S.; Park, S.H.; Heo, B.G.; Gorinstein, S. Extraction and characterization of some natural plant pigments. *Ind. Crop. Prod.* **2012**, *40*, 129–135. [CrossRef]
6. Borges, M.E.; Tejera, R.L.; Díaz, L.; Esparza, P.; Ibáñez, E. Natural dyes extraction from cochineal (*Dactylopius coccus*). New extraction methods. *Food Chem.* **2012**, *132*, 1855–1860. [CrossRef]
7. Tuli, H.S.; Chaudhary, P.; Beniwal, V.; Sharma, A.K. Microbial pigments as natural color sources: Current trends and future perspectives. *J. Food Sci. Technol.* **2015**, *52*, 4669–4678. [CrossRef] [PubMed]
8. Dufosse, L.; Fouillaud, M.; Caro, Y.; Mapari, S.A.; Sutthiwong, N. Filamentous fungi are large-scale producers of pigments and colorants for the food industry. *Curr. Opin. Biotechnol.* **2014**, *26*, 56–61. [CrossRef] [PubMed]
9. Feng, Y.; Shao, Y.; Chen, F. Monascus pigments. *Appl. Microbiol. Biotechnol.* **2012**, *96*, 1421–1440. [CrossRef] [PubMed]
10. Hajjaj, H.; Blanc, P.; Groussac, E. Kinetic analysis of red pigment and citrinin production by *Monascus ruber* as a function of organic acid accumulation. *Enzym. Microb. Technol.* **2000**, *27*, 619–625. [CrossRef]
11. Morales-Oyervides, L.; Oliveira, J.C.; Sousa-Gallagher, M.J.; Méndez-Zavala, A.; Montañez, J.C. Selection of best conditions of inoculum preparation for optimum performance of the pigment production process by *Talaromyces* spp. using the Taguchi method. *Biotechnol. Prog.* **2017**. [CrossRef] [PubMed]
12. Santos-Ebinuma, V.C.; Roberto, I.C.; Simas Teixeira, M.F.; Pessoa, A. Improving of red colorants production by a new *Penicillium purpurogenum* strain in submerged culture and the effect of different parameters in their stability. *Biotechnol. Prog.* **2013**, *29*, 778–785. [CrossRef] [PubMed]
13. Morales-Oyervides, L.; Oliveira, J.C.; Sousa-Gallagher, M.J.; Méndez-Zavala, A.; Montañez, J.C. Effect of heat exposure on the colour intensity of red pigments produced by *Penicillium purpurogenum* GH2. *J. Food Eng.* **2015**, *164*, 21–29. [CrossRef]
14. Dhale, M.A.; Vijay-Raj, A.S. Pigment and amylase production in *Penicillium* sp NIOM-02 and its radical scavenging activity. *Int. J. Food Sci. Technol.* **2009**, *44*, 2424–2430. [CrossRef]
15. Teixeira, M.; Teixeira, M.F.; Martins, M.; Caldas, J.; Kirsh, L.; Fernandes, O.; Fernandes, O.C.; Carneiro, A.; De Conti, R.; Durán, N. Amazonian biodiversity: Pigments from *Aspergillus* and *Penicillium*-characterizations, antibacterial activities and their toxicities. *Curr. Trends Biotechnol. Pharm.* **2012**, *6*, 300–311.
16. Sopandi, T.; Wardah, W. Sub-Acute toxicity of pigment derived from *Penicillium resticulosum* in mice. *Microbiol. Indones.* **2012**, *6*, 35–41. [CrossRef]
17. Ziolkowska, J.R.; Simon, L. Recent developments and prospects for algae-based fuels in the US. *Renew. Sustain. Energy Rev.* **2014**, *29*, 847–853. [CrossRef]
18. Glembin, P.; Kerner, M.; Smirnova, I. Cloud point extraction of microalgae cultures. *Sep. Purif. Technol.* **2013**, *103*, 21–27. [CrossRef]
19. Qureshi, N.; Maddox, I.S. Reduction in butanol inhibition by perstraction, utilization of concentrated Lactose/Whey permeate by *Clostridium acetobutylicum* to enhance butanol fermentation Economics. *Food Bioprod. Process.* **2005**, *83*, 43–52. [CrossRef]
20. Hu, Z.; Zhang, X.; Wu, Z.; Qi, H.; Wang, Z. Perstraction of intracellular pigments by submerged cultivation of *Monascus* in nonionic surfactant micelle aqueous solution. *Appl. Microbiol. Biotechnol.* **2012**, *94*, 81–89. [CrossRef] [PubMed]
21. Zhang, J.; Wang, Y.-L.; Lu, L.-P.; Zhang, B.-B.; Xu, G.-R. Enhanced production of Monacolin K by addition of precursors and surfactants in submerged fermentation of *Monascus purpureus* 9901. *Biotechnol. Appl. Biochem.* **2014**, *61*, 202–207. [CrossRef] [PubMed]
22. Wang, Y.; Zhang, B.; Lu, L.; Huang, Y.; Xu, G. Enhanced production of pigments by addition of surfactants in submerged fermentation of *Monascus purpureus* H1102. *J. Sci. Food Agric.* **2013**, *93*, 3339–3344. [CrossRef] [PubMed]

23. Wang, Z. The potential of cloud point system as a novel two-phase partitioning system for biotransformation. *Appl. Microbiol. Biotechnol.* **2007**, *75*, 1–10. [CrossRef] [PubMed]

24. Pan, T.; Wang, Z.; Xu, J.H.; Wu, Z.; Qi, H. Extractive fermentation in cloud point system for lipase production by *Serratia marcescens* ECU1010. *Appl. Microbiol. Biotechnol.* **2010**, *85*, 1789–1796. [CrossRef] [PubMed]

25. Zhang, W.; Wang, Z.; Li, W.; Zhuang, B.; Qi, H. Production of L-phenylacetylcarbinol by microbial transformation in polyethylene glycol-induced cloud point system. *Appl. Microbiol. Biotechnol.* **2008**, *78*, 233–239. [CrossRef] [PubMed]

26. Kang, B.; Zhang, X.; Wu, Z.; Qi, H.; Wang, Z. Effect of pH and nonionic surfactant on profile of intracellular and extracellular *Monascus* pigments. *Process Biochem.* **2013**, *48*, 759–767. [CrossRef]

27. Cruz-Hernández, M; Rodráguez, R. Aislamiento y caracterización morfológica de cepas microbianas degradadotas de taninos. In Proceedings of the XXII Annual Meeting of the Mexican Academy of Chemical Engineering (AMIDIQ), Mazatlan, Mexico, 9 September 2001; pp. 71–72.

28. Espinoza-Hernández, T. Caracterización Morfológica, Fisiológica y Molecular de Tres Cepas Fúngicas Productoras de Pigmentos. Bachelor's Thesis, Universidad Autónoma de Coahuila, Saltillo, Mexico, 2004.

29. Frisvad, J.C.; Yilmaz, N.; Thrane, U.; Rasmussen, K.B.; Houbraken, J.; Samson, R.A. Talaromyces atroroseus, a new species efficiently producing industrially relevant red pigments. *PLoS ONE* **2013**, *8*, e84102. [CrossRef] [PubMed]

30. Méndez-Zavala, A. Identificación de Factores que Afectan la Producción de Pigmentos por *Penicillium purpurogenum* GH2 y Obtención de Perfiles Cromatográficos. Master's Thesis, Universidad Autónoma de Coahuila, Saltillo, Mexico, 2011.

31. Morales-Oyervides, L.; Oliveira, J.C.; Sousa-Gallagher, M.J.; Méndez-Zavala, A.; Montañez, J.C. Quantitative assessment of the impact of the type of inoculum on the kinetics of cell growth, substrate consumption and pigment productivity by *Penicillium purpurogenum* GH2 in liquid culture with an integrated stochastic approach. *Food Bioprod. Process.* **2015**, *96*, 221–231. [CrossRef]

32. Méndez-Zavala, A.; Pérez, C.; Montañez, J.C.; Martínez, G.; Aguilar, C.N. Red pigment production by *Penicillium purpurogenum* GH2 is influenced by pH and temperature. *J. Zhejiang Univ. Sci. B* **2011**, *12*, 961–968. [CrossRef] [PubMed]

33. Dubois, M.; Gilles, K.; Hamilton, J. Colorimetric method for determination of sugars and related substances. *Anal. Chem.* **1956**, *28*, 300–306. [CrossRef]

34. Kang, B.; Zhang, X.; Wu, Z.; Qi, H.; Wang, Z. Isolation of natural red colorants from fermented broth using ionic liquid-based aqueous two-phase systems. *J. Ind. Microbiol. Biotechnol.* **2013**, *40*, 507–516.

35. Babitha, S.; Soccol, C.R.; Pandey, A. Effect of stress on growth, pigment production and morphology of *Monascus* sp. in solid cultures. *J. Basic Microbiol.* **2007**, *47*, 118–126. [CrossRef]

36. Velmurugan, P. Studies on the Production and Dyeing Properties of Water Soluble Pigments from Filamentous Fungi. Ph.D. Thesis, Bharathiar University, Coimbatore, India, 2008.

37. Dafoe-Samper, T. Properties of Absorbent Polymer Extractants for the Selective Removal of Target Molecules from Fermentation Systems. Ph.D. Thesis, Queen's University, Kingston, ON, Canada, 2014.

38. Kim, S.W.; Seo, W.T.; Park, Y.H. Enhanced production of β-carotene from *Blakeslea trispora* with Span 20. *Biotechnol. Lett.* **1997**, *19*, 557–560. [CrossRef]

39. Zhang, H.; Xia, Y.J.; Wang, Y.L.; Zhang, B.B.; Xu, G.R. Coupling use of surfactant and in situ extractant for enhanced production of Antrodin C by submerged fermentation of *Antrodia camphorata*. *Biochem. Eng. J.* **2013**, *79*, 194–199. [CrossRef]

40. Hu, Z.; Zhang, X.; Wu, Z.; Qi, H.; Wang, Z. Export of intracellular *Monascus* pigments by two-stage microbial fermentation in nonionic surfactant micelle aqueous solution. *J. Biotechnol.* **2012**, *162*, 202–209. [CrossRef] [PubMed]

41. Dhamole, P.B.; Wang, Z.; Liu, Y.; Wang, B.; Feng, H. Extractive fermentation with non-ionic surfactants to enhance butanol production. *Biomass Bioenergy* **2012**, *40*, 112–119. [CrossRef]

42. Deive, F.J.; Carvalho, E.; Pastrana, L.; Rúa, M.L.; Longo, M.A.; Sanroman, M.A. Strategies for improving extracellular lipolytic enzyme production by *Thermus thermophilus* HB27. *Bioresour. Technol.* **2009**, *100*, 3630–3637. [CrossRef] [PubMed]

43. Mavituna, F.; Sinclair, C. *Practical Fermentation Technology*; John Wiley & Sons: West Sussex, UK, 2008; p. 402.

44. Van Bodegom, P. Microbial maintenance: A critical review on its quantification. *Microb. Ecol.* **2007**, *53*, 513–523. [CrossRef] [PubMed]
45. Patil, S.; Sivanandhan, G.; Thakare, D. Effect of physical and chemical parameters on the production of Red exopigment from *Penicillium purpurogenum* isolated from spoilt onion and study of its antimicrobial activity. *Int. J. Curr. Microbiol. Appl. Sci.* **2015**, *4*, 599–609.

Journal of
Fungi

MDPI

Communication

Utilization of High Performance Liquid Chromatography Coupled to Tandem Mass Spectrometry for Characterization of 8-*O*-methylbostrycoidin Production by Species of the Fungus *Fusarium*

Mark Busman

Mycotoxin Prevention and Applied Microbiology Research Unit, National Center for Agricultural Utilization Research, Agricultural Research Service, United States Department of Agriculture, Peoria, IL 61604, USA; Mark.Busman@ars.usda.gov; Tel.: +1-309-681-6241

Received: 1 June 2017; Accepted: 20 July 2017; Published: 25 July 2017

Abstract: The pigment 8-*O*-methylbostrycoidin is a polyketide metabolite produced by multiple species of the fungus *Fusarium* that infects plant crops, including maize. A technique was developed for the analysis of 8-*O*-methylbostrycoidin by high performance liquid chromatography coupled to electrospray ionization tandem mass spectrometry. The quantitative nature of the LC-MS/MS experiment was demonstrated over a range of concentrations in maize. Limits of detection for the method (10 ng/g from 8-*O*-methylbostrycoidin spiked into ground maize) were shown, and susceptibility of the method to matrix effects from maize was also evaluated. The method was applied to evaluate the ability of the maize pathogen *Fusarium verticillioides* to produce 8-*O*-methylbostrycoidin in developing maize ears grown in an agricultural field.

Keywords: maize; liquid chromatography-mass spectrometry; 8-*O*-methylbostrycoidin; pigment

---

1. Introduction

The pigment, 8-*O*-methylbostrycoidin (Figure 1), is an azaanthraquinone secondary metabolite that is widely produced by species of the filamentous fungus *Fusarium*. A complete structural characterization was accomplished by Steyn [1]. Total synthesis of the compound was demonstrated by Cameron [2,3]. The compound was first observed from *Fusarium* isolates taken from maize [1] and citrus [4]. Deng has recently noted production from a marine *Aspergillus* isolate [5].

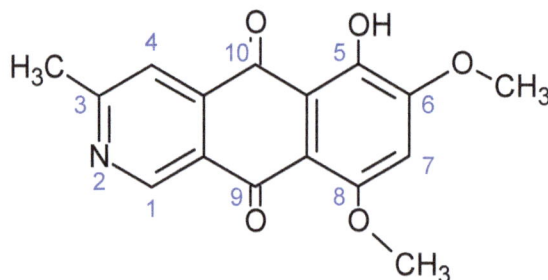

Figure 1. Structure of 8-*O*-methylbostrycoidin.

Early studies described the isolation of 8-*O*-methylbostrycoidin and showed its cytotoxicity against several mammalian cell lines, including rat hepatoma, Madin Darby canine kidney and Chinese hamster ovary [6]. Later, Hashimoto again observed the low cytotoxicity against the rat hepatoma cell line, but observed a strong ability to inhibit hepatic glucose production [7]. More recently, 8-*O*-methylbostrycoidin has been evaluated for cytotoxicity to a number of human cancer cell lines and has been shown to strongly inhibit α-acetylcholine esterase [5].

It is thought that 8-*O*-methylbostrycoidin biosynthesis involves a polyketide synthase that catalyzes the synthesis of a heptaketide, the first intermediate in 8-*O*-methylbostrycoidin biosynthesis. It has been suggested that a common polyketide synthase is responsible for the initial stages in the synthesis of several anthraquinones produced by *Fusaria* [8–10].

Chemical characterization of 8-*O*-methylbostrycoidin frequently is based upon its deep red color. For example, 8-*O*-methylbostrycoidin production in fungi grown on culture plates can be monitored visually, based on the color of the substrate. Levels of 8-*O*-methylbostrycoidin have also been monitored by absorbance at 510–520 nm either in HPLC detection or in spectrophotometric analysis of *Fusarium* extracts. Such UV-VIS based detection of related compounds has been implicated as an interfering actor in the detection of another Fusarium metabolite, zearalanone [11].

Here, we demonstrate a convenient technique for the analysis of 8-*O*-methylbostrycoidin by high performance liquid chromatography coupled to electrospray ionization tandem mass spectrometry. The technique is applied to the screening of *Fusarium* species for 8-*O*-methylbostrycoidin production on cracked maize substrate and to the evaluation of field production by *F. verticillioides* in maize.

2. Materials and Methods

2.1. Materials

Unless otherwise noted, all solvents were of HPLC grade from Sigma Chemical Co. (St. Louis, MO, USA). Water was from a Millipore (Billerica, MA, USA) water purification system. Purified 8-*O*-methylbostrycoidin was obtained from cultures of *F. verticillioides* as previously described [6]. Mature cultures of *F. verticillioides* MRC-826 on cracked maize were extracted with 60% aqueous methanol and purified on with successive Amberlite XAD-2 (Sigma, St. Louis, MO, USA) and normal phase (silica gel) liquid chromatographies. Eluted 8-*O*-methylbostrycoidin was monitored by ESI-MS. Purified 8-*O*-methylbostrycoidin was evaluated by ^{1}H and ^{13}C NMR. A stock solution of 1 mg/mL 8-*O*-methylbostrycoidin in acetonitrile was used for constructing calibration solutions and for spike solutions.

2.2. Fungal Strains and Preparation of Maize Materia

Two strains of *F. verticillioides* isolates were obtained from field-grown maize kernels as previously described [12]. Briefly, maize kernels were surface sterilized by soaking in a 10% chlorine bleach solution for 1 min. After rinsing three times in sterile water, the kernels were placed on Nash agar medium [12] and incubated for 3–5 days in the dark at room temperature. Single-spore derived isolates of the resulting fungal growth were obtained as described previously [13]. The resulting colonies were identified as *F. verticillioides* first by morphology in comparison with morphological descriptions of the fungus [13] and then by DNA sequence analysis of the gene encoding the translation elongation factor 1-α and comparison of the resulting sequence data to the Fusarium ID Database [14]. Cracked maize kernel cultures of *Fusarium* were prepared as previously described [15].

2.3. Extraction of 8-*O*-methylbostrycoidin from Maize

Twenty grams of cracked or ground maize were accurately weighed and extracted with 50 mL of acetonitrile for 2 h on a Gyratory Model G2 Shaker (New Brunswick Scientific, Edison, NJ, USA). Slurries of sample with solvent were centrifuged for 5 min at ~1500× *g*. One milliliter aliquots of the centrifugate were decanted by pipette into sample vials for LC-MS analysis.

2.4. ESI-MS and MS/MS

Unless otherwise noted, all experiments were conducted utilizing a ThermoFinnigan LCQ-DECA (Thermo Scientific, San Jose, CA, USA) ion trap mass spectrometer equipped with an electrospray ionization source. Flow injection experiments were accomplished with by injecting 10 µL plugs of analyte solution into a 300 µL/min flow of 1/1 methanol-water with 0.3% acetic acid. Injection of the analyte plugs was done by use of a Rheodyne Model 7125 injector (IDEX, Oak Harbor, WA, USA) valve fitted with a 10 µL injection loop.

2.5. HPLC-MS/MS

Ten microliters of extract analyzed with a MetaChem (Torrance, CA, USA) Inertsil C18 (150 mm length, 3 mm diameter, 5 µm diameter particle, 100 Å pore size) column. Chromatography was performed utilizing a ThermoSpectraPhysics (Thermo Scientific, San Jose, CA, USA) high performance liquid chromatography system consisting of an AS4000 autosampler coupled to a P2000 gradient pump. Elution of analyte was achieved with a 300 µL/min gradient flow of methanol and water (0.3% acetic acid was added to the mobile phase.) The solvent program used a 35–95% gradient over 25 min. The mass spectrometer was operated in positive mode electrospray ionization mode with an ionization voltage of 4.5 kV. The inlet capillary temperature was 255 °C, the inlet capillary voltage was 46 V and the tube lens offset was −5 V. The collision energy was optimized to 33%. Operation of the chromatography and mass spectrometry instrument and quantitation of the eluting 8-*O*-methylbostrycoidin was done utilizing ThermoFinnigan Xcalibur software (Version 1.4, Thermo Scientific, San Jose, CA, USA). Quantitative data was processed using Xcalibur QualBrowser (ThermoFinnigan) software. Uncertainties in mean determinations are expressed in terms of standard deviation.

Calibration curves were plotted using pure standards as well as maize matrix-assisted standards (acetonitrile extract). Maize matrix assisted standards were prepared as follows: Maize extract (considered blank for 8-*O*-methylbostrycoidin), which was obtained as described above, was spiked with 8-*O*-methylbostrycoidin dissolved at an appropriate concentration in acetonitrile.

2.6. Spike/Recovery Experiments and Method Validation

Spike recovery experiments were carried out in order to validate the method. Spiking solutions were prepared in acetonitrile and appropriate amounts added to cracked maize 1 h before extraction. Maize samples were spiked with 8-*O*-methylbostrycoidin by adding 250 µL aliquots (0.5, 5 and 50 µg/mL) of 8-*O*-methylbostrycoidin. Spiked samples were extracted with the same methods used for other samples. Spike/recovery samples were analyzed in triplicate on three successive days to evaluate method stability. Quantitations were based on integration of chromatographic peak area corresponding to the detection of characteristic fragment ions arising from the collisionally induced dissociation of the parent pseudomolecular ion $[M + H]^+$ of the analyte. Analyte responses were compared to responses obtained from solubilized (acetonitrile) 8-*O*-methylbostrycoidin. 8-*O*-methylbostrycoidin was quantified using an external calibration curve, which was constructed by plotting the concentration against the signal area.

2.7. Field Test

Maize line B73 was grown in field plots at the USDA, ARS, NCAUR (Peoria County, IL, USA) following guidelines of the Animal and Plant Health Inspection Service (APHIS) outlined in APHIS permit number P526-09-01825. The resulting maize ears were infected with the *F. verticillioides* by injection of 2 mL of a solution of 1×10^6 *F. verticillioides* spores per mL water into the silk channel of each ear at 4 to 6 days after silk emergence [16]. Control ears were similarly injected with water. For each treatment 10 ears were inoculated. Upon maturity, ears were harvested and visually evaluated for extent of fungal infestation as described by Desjardins et al. [17]. The disease evaluation method provides a 0–7 scoring of the damage of the kernels of the ear where a higher score indicates greater damage and a lower score indicates lesser damage. The ears were shelled and the resulting kernels were ground with a Stein Model M2 Laboratory Mill (Steinlite Corporation, Atchison, KS, USA). Ground kernels for each ear were extracted for LC-MS/MS analysis.

3. Results

3.1. MS and MS/MS by Flow Injection

The ESI-MS spectrum for the flow injection of a solution of 8-O-methylbostrycoidin is shown in Figure 2a. Operating the ion-trap mass analyzer in the MS/MS mode has potential to increase the selectivity of the method, minimize background signal and improve method performance. To further assess 8-O-methylbostrycoidin mass spectrometric behavior, tandem mass spectra for the collisional induced dissociation of the $[M + H]^+$ peak were acquired. When a certain threshold collision energy level was exceeded, the molecule broke up into several fragments. Features of the spectra did not change substantially despite use of higher collision energies. The resulting spectra were extremely simple. A spectrum for a CID-MS/MS experiment is shown in Figure 2b (33% collision energy). The instrument was operated to allow selection of a parent ion (m/z 300) and scanning of product ions. The dominant fragment ions at m/z 241, 256 and 271 are likely the result of facile losses of acetic acid, carbon dioxide and methanol. Other minor fragments are likely the result of extensive rearrangements. The molecular ion shows a great deal of stability despite use of relatively high collision energy. Lower mass fragments resulting from a more major decomposition of the parent ion would likely fall beneath the lower mass limit of the MS detector during the MS/MS experiment.

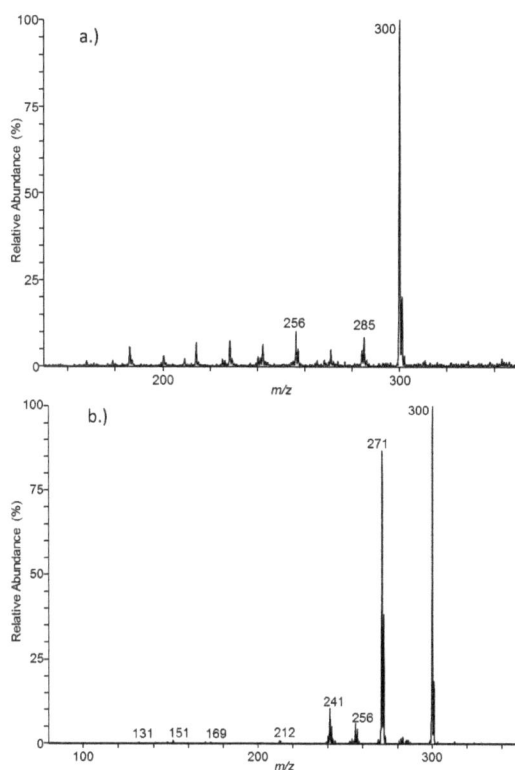

Figure 2. (a) MS and (b) MS/MS spectra for 8-O-methylbostrycoidin.

3.2. HPLC Coupled to MS

Injections of 8-*O*-methylbostrycoidin (10 μL, 50 μg/mL) were made into reversed phase gradient flows of methanol and water on the HPLC column. Good chromatographic behavior was observed with 10 μL injections of concentrations up to 1000 μg/mL. Figure 3 shows a LC-MS/MS profile resulting from of a 50 μg/mL solvent standard injection using multiple reaction monitoring (MRM) reflecting detection of the three fragments and intact parent $[M + H]^+$ ion.

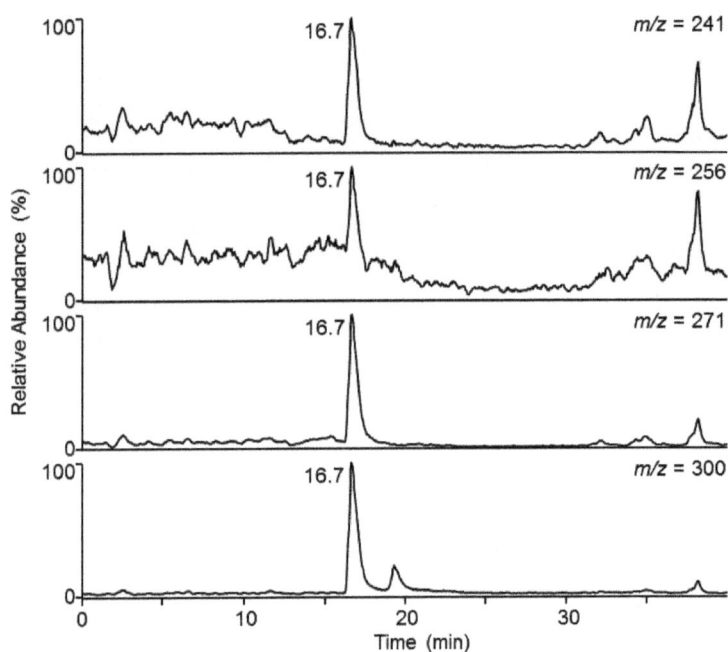

Figure 3. Chromatogram for elution of a 50 μg/mL solvent standard of 8-*O*-methylbostrycoidin.

3.3. LC-MS/MS Quantitation

For quantitative analyses, abundant fragments from the $[M + H]^+$ ion of the analyte, 8-*O*-methylbostrycoidin, were monitored in the MRM mode. Signals from the fragments were maximized by optimizing collision energy. To assess quantitative behavior of the LC-MS system, 10 μL injections over a wide range of 8-*O*-methylbostrycoidin concentrations were made. Response curves were obtained over a range of 8-*O*-methylbostrycoidin concentrations. Based on the observed response, it was determined that the monitoring of the m/z 271 fragment ion provided a good basis for quantitation of 8-*O*-methylbostrycoidin. Figure 4 shows an example plot of integrated response against concentration for the m/z 271 fragment ion from solvent and matrix standards. Good linearity, up to 1000 μg/mL, is observed. Based on the observation of the transition to m/z 271 ion, the absolute minimum detection limit (based on the signal-to-noise ratio of 3) was better than 3 ng/g from 8-*O*-methylbostrycoidin spiked into ground maize. Further, the limit of quantitation was 10 ng/g from 8-*O*-methylbostrycoidin spiked into ground maize.

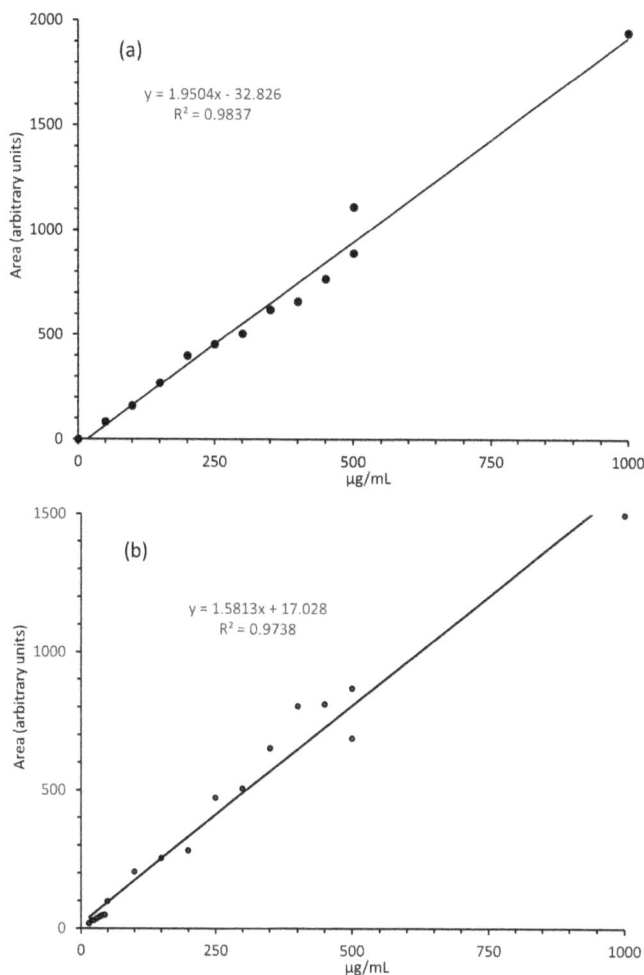

Figure 4. Calibration curves for 8-O-methylbostrycoidin (**a**) solvent and (**b**) matrix standards.

3.4. Spike Recovery Studies

The goal of this work was to have a convenient method for determination of 8-O-methylbostrycoidin from maize substrate. To assess the ability of the LC-MS/MS technique to serve as the basis for such a method, spike-recovery experiments were conducted. A series of portions of cracked maize were spiked with 8-O-methylbostrycoidin (acetonitrile) in order to achieve levels corresponding to 0.0125, 0.125 and 1.25 µg/g 8-O-methylbostrycoidin. The samples were then extracted and processed according to the above described method. The analysis of the spiked samples ($n = 9$ for each spike level) yielded recoveries (72.8 ± 35.6, 0.0125 µg/g; 41.3 ± 17.5, 0.125 µg/g; 59.9 ± 32.3, 1.25 µg/g) of the 8-O-methylbostrycoidin upon quantitation utilizing the m/z 271 fragment ion.

3.5. Matrix Interference

Experiments to evaluate matrix effects were in conducted according to guidelines described by Matuszewski et al. [18]. LC-MS/MS areas of standards formulated in solvent (acetonitrile) were

J. Fungi **2017**, *3*, 43

compared with those measured in a blank cracked maize extract spiked, after extraction, with the same analyte amounts. Tests were conducted on spiked maize extract samples (8-*O*-methylbostrycoidin concentration of 0.5 µg/mL). The level of 8-*O*-methylbostrycoidin in the maize extract used for the spiking experiments was assumed to be zero. The average of measured matrix effect (%) from solvent and extract samples run in pairs (*n* = 8) was 154 ± 12%. In the case of maize extract, the matrix actually provides an enhancement to the response of the LC-MS/MS detection.

3.6. Determination of 8-O-methylbostrycoidin from Field Grown Maize

Field tests were conducted to evaluate the ability of *F. verticillioides* to produce 8-*O*-methylbostrycoidin in maize kernels under agriculturally relevant conditions. Injection of suspensions of spores from two different strains of *F. verticillioides* into developing maize ears resulted in relatively high levels of maize ear rot symptoms at harvest (Table 1). Injection of the different strains of *F. verticillioides* resulted in markedly different levels of 8-*O*-methylbostrycoidin in the resulting kernels (Table 1). Control injection of ears with water resulted in low levels of disease symptoms, as previous reported [17], and 8-*O*-methylbostrycoidin was not detected in kernels harvested from control-injected ears.

Table 1. Results for field evaluation of disease level and 8-*O*-methylbostrycoidin production in maize ears upon inoculation with *F. verticillioides* isolates.

Strain	Disease Score [1]	Level of 8-*O*-methylbostrycoidin (µg/g) [1]
AMR 5	4.9 (1.9)	0.5 (0.9)
AMR 10	4.4 (1.6)	2.6 (4.8)
Control inoculation	1.3 (0.9)	0 (–)

[1] Uncertainties in determinations in terms of standard deviations are indicated in parentheses. Each maize ear was considered a repetition of the inoculation experiment *n* = 10. No standard deviation was calculated for the levels of the analyte detected in the control inoculation kernels. This is indicated by the (–).

4. Discussion

Currently, common methods for the evaluation of 8-*O*-methylbostrycoidin production by fungi are based upon UV absorbance. Potentially, such methods would be susceptible to interferences from extracted absorbing compounds produced by maize, as well as invading fungi. The method described here offers the potential of advantages in sensitivity and selectivity associated with LC-MS/MS. Our examination of the performance of the method provides us with an indication of its sensitivity, as well as its tolerance of matrix effects from the maize. The developed method allows the convenient determination of 8-*O*-methylbostrycoidin without a substantial effort for sample clean-up. As far as we are aware, this is the first report of production of 8-*O*-methylbostrycoidin in field-grown maize.

Acknowledgments: The author wishes to acknowledge the technical contributions by Debbie Shane.

Author Contributions: Mark Busman conceived and designed the experiments, performed the experiments, analyzed the data and wrote the paper.

Conflicts of Interest: The author declares no conflict of interest. Mention of trade names or commercial products in this article is solely for the purpose of providing specific information and does not imply recommendation or endorsement by the U.S. Department of Agriculture.

References

1. Steyn, P.S.; Wessels, P.L.; Marasas, W.F.O. Pigments from *Fusarium moniliforme* sheldon. Structure and ^{13}C nuclear magnetic resonance assignments of an azaanthraquinone and three naphthoquinones. *Tetrahedron* **1979**, *35*, 1551–1555. [CrossRef]
2. Cameron, D.W.; Deutscher, K.R.; Feutrill, G.I. Synthesis of bostrycoidin and 8-*O*-methlylbostrycoidin. *Tetrahedron Lett.* **1980**, *21*, 5089–5090. [CrossRef]

3. Cameron, D.W.; Deutscher, K.R.; Feutrill, G.I. Nucleophilic alkenes. Ix. Addition of 1,1-dimethoxyethene to azanaphthoquinones: Synthesis of bostrycoidin and 8-*O*-methylbostrycoidin. *Aust. J. Chem.* **1982**, *35*, 1439. [CrossRef]

4. Tatum, J.H.; Baker, R.A.; Berry, R.E. Naphthoquinones produced by *Fusarium oxysporum* isolated from citrus. *Phytochemistry* **1985**, *24*, 457–459. [CrossRef]

5. Deng, C.M.; Liu, S.X.; Huang, C.H.; Pang, J.Y.; Lin, Y.C. Secondary metabolites of a mangrove endophytic fungus *Aspergillus terreus* (no. Gx7-3b) from the South China Sea. *Mar. Drugs* **2013**, *11*, 2616–2624. [CrossRef] [PubMed]

6. Vesonder, R.F.; Gasdorf, H.; Peterson, R.E. Comparison of the cytotoxicities of *Fusarium* metabolites and *Alternaria* metabolite AAL-toxin to cultured mammalian cell lines. *Arch. Environ. Contam. Toxicol.* **1993**, *24*, 473–477. [CrossRef] [PubMed]

7. Hashimoto, J.; Motohashi, K.; Sakamoto, K.; Hashimoto, S.; Yamanouchi, M.; Tanaka, H.; Takahashi, T.; Takagi, M.; Shin-ya, K. Screening and evaluation of new inhibitors of hepatic glucose production. *J. Antibiot. (Tokyo)* **2009**, *62*, 625–629. [CrossRef] [PubMed]

8. Awakawa, T.; Kaji, T.; Wakimoto, T.; Abe, I. A heptaketide naphthaldehyde produced by a polyketide synthase from *Nectria haematococca. Bioorg. Med. Chem. Lett.* **2012**, *22*, 4338–4340. [CrossRef] [PubMed]

9. Brown, D.W.; Butchko, R.A.E.; Busman, M.; Proctor, R.H. Identification of gene clusters associated with fusaric acid, fusarin, and perithecial pigment production in *Fusarium verticillioides. Fungal Genet. Biol.* **2012**, *49*, 521–532. [CrossRef] [PubMed]

10. Studt, L.; Wiemann, P.; Kleigrewe, K.; Humpf, H.U.; Tudzynski, B. Biosynthesis of fusarubins accounts for pigmentation of *Fusarium fujikuroi* perithecia. *Appl. Environ. Microbiol.* **2012**, *78*, 4468–4480. [CrossRef] [PubMed]

11. Smith, J.S.; Fotso, J.; Leslie, J.F.; Wu, Y.; VanderVelde, D.; Thakur, R.A. Characterization of bostrycoidin: An analytical analog of zearalenone. *J. Food Sci.* **2004**, *69*, C227–C232. [CrossRef]

12. Nelson, P.E.; Desjardins, A.E.; Plattner, R.D. Fumonisins, mycotoxins produced by *Fusarium* species: Biology, chemistry, and significance. *Annu. Rev. Phytopathol.* **1993**, *31*, 233–252. [CrossRef] [PubMed]

13. Leslie, J.F.; Summerell, B.A. *The Fusarium Laboratory Manual*; Blackwell Publishing: Ames, IA, USA, 2006.

14. Geiser, D.M.; Jimenez-Gasco, M.D.M.; Kang, S.; Makalowska, I.; Veeraraghavan, N.; Ward, T.J.; Zhang, N.; Kuldau, G.A.; O'Donnell, K. Fusarium-ID v. 1.0: A DNA sequence database for identifying *Fusarium. Eur. J. Plant Pathol.* **2004**, *110*, 473–480. [CrossRef]

15. Seo, J.A.; Proctor, R.H.; Plattner, R.D. Characterization of four clustered and coregulated genes associated with fumonisin biosynthesis in *Fusarium verticillioides. Fungal Genet. Biol.* **2001**, *34*, 155–165. [CrossRef] [PubMed]

16. Reid, L.; Hamilton, R.; Mather, D. Screening maize for resistance to *gibberella* ear rot. In *Technical Bulletin 1996-5e*; Agriculture & Agri-food Canada: Ottawa, ON, Canada, 1996; pp. 1–40.

17. Desjardins, A.E.; Plattner, R.D. Fumonisin B1-nonproducing strains of *Fusarium verticillioides* cause maize (*Zea mays*) ear infection and ear rot. *J. Agric. Food Chem.* **2000**, *48*, 5773–5780. [CrossRef] [PubMed]

18. Matuszewski, B.K.; Constanzer, M.L.; Chavez-Eng, C.M. Strategies for the assessment of matrix effect in quantitative bioanalytical methods based on HPLC-MS/MS. *Anal. Chem.* **2003**, *75*, 3019–3030. [CrossRef] [PubMed]

Journal of *Fungi*

|MDPI|

Article

Assessment of the Dyeing Properties of the Pigments Produced by *Talaromyces* spp.

Lourdes Morales-Oyervides [1,2], Jorge Oliveira [1], Maria Sousa-Gallagher [1], Alejandro Méndez-Zavala [2] and Julio Cesar Montañez [2,*]

[1] School of Engineering, University College Cork, Cork, Ireland; lourdesmorales@uadec.edu.mx (L.M.-O.); j.oliveira@ucc.ie (J.O.); M.deSousaGallagher@ucc.ie (M.S.-G.)

[2] Department of Chemical Engineering, Universidad Autónoma de Coahuila. Saltillo, Coahuila 25280, Mexico; alejandro.mendez@uadec.edu.mx

* Correspondence: julio.montanez@uadec.edu.mx; Tel.: +52-844-416-9213

Received: 1 June 2017; Accepted: 28 June 2017; Published: 5 July 2017

Abstract: The high production yields of pigments by *Talaromyces* spp. and their high thermal stability have implied that industrial application interests may emerge in the food and textile industries, as they both involve subjecting the colourants to high temperatures. The present study aimed to assess the potential application of the pigments produced by *Talaromyces* spp. in the textile area by studying their dyeing properties. Dyeing studies were performed on wool. The dyeing process consisted of three stages: scouring, mordanting, and dyeing. Two different mordants (alum, A; ferric chloride, F) were tested at different concentrations on fabric weight (A: 5, 10, 15%; F: 10, 20, 30%). The mordanting process had a significant effect on the final colour of the dyed fabrics obtained. The values of dyeing rate constant (k), half-time of dyeing ($t_{1/2}$), and sorption kinetics behaviour were evaluated and discussed. The obtained results showed that pigments produced by *Talaromyces* spp. could serve as a source for the natural dyeing of wool textiles.

Keywords: fungal pigments; dyeing properties; textiles; *Talaromyces*

1. Introduction

Market trends to use natural colourants as food additives [1], natural dyes [2], functional foods [3], and cosmetic products [4] represent an opportunity for the application of natural pigments in several sectors of industry.

The possibility to exploit biological sources such as microorganisms for the production of natural pigments has been recommended [5]. However, the successful application of microbial pigments relies on high production yields, reasonable production costs and capital investment, regulatory approval, pigment characterisation, and stability to environmental factors such as temperature and light. Among the microorganisms with potential to produce a vast variety of pigments (*Monascus* homologues) is *Talaromyces* spp. (formerly *Penicillium* spp.) [6,7].

There has been a considerable effort in performing studies regarding the optimisation of the production of *Talaromyces* pigments [8–11]; however, there are only a few studies on the application of pigments produced by this strain [12,13].

Talaromyces pigments are thermally stable [14], implying that industrial application interests can emerge in the food and textile industries, as both processes involve subjecting the colourants to high temperatures.

Colourants used by the textile industry are predominantly synthetic; however, synthetic dyes are not environment-friendly, and recently the textile industry has been challenged to ensure compliance with environmental issues [15]. Natural colourants are environmentally friendly and biodegrade more quickly than synthetic dyes [16].

Additionally, natural pigments present properties of great interest in the textile industry, such as antibacterial properties. Fabrics can act as carriers of bacteria responsible for undesirable odours. It has been shown that *Talaromyces* pigment extracts possess antimicrobial properties [17]. These properties, along with the absence of toxicity [18,19], make them a valuable alternative as natural colourants in the textile industry.

However, natural colourants still face a huge disadvantage against synthetic colourants. Dyeing textile with natural colourants usually involves issues of limited shade range and lower fastness properties of the dyed fabrics. These problems have been overcome applying a pretreatment to the textile with mordants in order to create affinity between the fibre and the dye. Selection of mordant and concentration is important in natural dyeing processes, as mordants can increase the depth of shade or drastically alter the final colour of the dyed fabric [2].

The present study aimed to assess the potential application of the pigments produced by *Talaromyces* spp. in the textile area by studying their dyeing capacity.

2. Materials and Methods

2.1. Microorganism and Inoculum Preparation

Talaromyces spp. was used for the production of red pigments (Department of Food Science and Technology, Autonomous University of Coahuila, Saltillo, Coahuila). The purified strain had been previously isolated and characterised as *Penicillium purpurogenum* GH2 [20,21]. *Penicillium purpurogenum* has, however, been transferred to *Talaromyces* spp. [22]. The strain was maintained on PDA (Potato dextrose agar) slants at 4 °C and sub-cultured periodically. Inoculum was prepared in Erlenmeyer flasks (125 mL in capacity) containing 25 mL of Potato Dextrose Broth medium (PDB medium, ATCC medium:336), which were sterilised and inoculated with a spore suspension (1×10^5 spores/mL) of *Talaromyces* spp. previously incubated for 5 days. The flasks were then incubated at 30 °C for 84 h in an orbital shaker (Innova 94, New Brunswick Scientific, Edison, NJ, USA) at 200 rpm [9,23].

2.2. Culture Media

The PDA medium was prepared with a concentration of 39.0 g/L (Bioxon, Mexico). The medium PDB medium was prepared by finely boiling 0.3 kg of diced potatoes in 500 mL of water until thoroughly cooked; then the potatoes were filtered through cheesecloth and water was added to the filtrate to complete a volume of 1.0 L. Finally, 20.0 g of glucose was added before sterilisation. The Czapek-dox modified medium [24] consisted of (g/L): D-xylose 15.0, $NaNO_3$ 3.0, $MgSO_4 \cdot 7H_2O$ 0.5, $FeSO_4 \cdot 7H_2O$ 0.1, K_2HPO_4 1.0, KCl 1.0 and ethanol 20.0.

2.3. Cultivation Conditions for Pigment Production

The initial pH of the Czapek-dox modified medium was adjusted to 5.0 before sterilising by using 0.22 μm sterile membranes (Millipore, Billerica, MA, USA). A mycelial suspension of *Talaromyces* spp. was inoculated at 10% (*v/v*) in 125 mL Erlenmeyer flasks containing 25 mL of medium. The inoculated flasks were incubated at 30 ± 2 °C in an orbital shaker (Innova 94, New Brunswick Scientific, Edison, NJ, USA) at 200 rpm for 6 days.

2.4. Pigment Extraction

The pigment extraction was performed according to the methodology reported by Méndez-Zavala et al. (2011) [25]. The pigment extract was centrifuged at 8000 rpm and at 4 °C for 20 min (Sorvall, Primo R Biofuge Centrifugation Thermo, Waltham, MA, USA) and then filtered through a 0.45 μm cellulose membranes (Millipore, Billerica, MA, USA) for the subsequent analysis of pigments. In this study, only extracellular pigments were considered. The analysis of red pigment production was conducted by measuring the absorbance of the filtered extract at 500 nm using a spectrophotometer (Cary 50, UV-Visible Varian, Palo Alto, CA, USA). This wavelength was selected by scanning the maximum sensitivity for the presence of the pigment (that is, the pigment absorbs maximum light at this wavelength). Red pigment extracts were stored in the dark at 4 °C before being used for the subsequent test.

2.5. Wool Dyeing Process

Wool fabric was bought at a local textile company. The dyeing process consisted of three stages: scouring [26], mordanting, and dyeing [12].

2.5.1. Fabric Scouring

The fabric was first scoured to remove any impurities so that they would not interfere with the dyeing process [26]. Wool was rinsed with 100 mL of hot distilled water (60 °C) per gram of wool; then, 18 mL of neutral soap per litre of water was added, along with 25% *w/v* of sodium carbonate, taking into consideration the wool weight. The mixture was heated to 60 °C and the wool was constantly moved from top to bottom. Then, the fabric was washed with water at ambient temperature (six times).

2.5.2. Fabric Mordanting

The fabric mordanting process was carried out using the pre-mordanting technique [12]. To assess the most appropriate mordant for *Talaromyces* spp. pigments, two different mordants were tested at different concentrations on the weight of fabric (Table 1). Concentrations were selected from literature according to recommended concentrations for these two mordants [27]. Wool was heated in mordant solutions at 70 °C for 1 h. Subsequently, the fabric was squeezed to remove excess liquid and then air dried at room temperature overnight.

Table 1. Mordants studied for the pre-mordanting process and selected concentrations. Codes were assigned.

Mordant	Concentration, % *w/w*	Code
Ferric chloride	10	F1
	20	F2
	30	F3
Alum	5	A1
	10	A2
	15	A3

2.6. Dyeing

The pre-mordanted wool was dyed with 40 mL of pigment extract (40:1, pigment extract per gramme of fabric) in a conical flask at 80 °C for 90 min. pH was not controlled. The colour of fabrics after dyeing was determinate by CIELAB colour coordinates using ColorEye XTS colorimeter (GretagMacbeth, Grand Rapids, MI, USA).

Pigment uptake was determined by measuring the optical density of the dye solution samples at a wavelength of 500 nm [27]. Percentage of pigment uptake (q, %) was calculated using the following equation:

$$q = \frac{OD_o - OD_i}{OD_o} * 100 \tag{1}$$

where OD_o is the initial optical density (500 nm) of the dye bath, and OD_i is the optical density after dyeing at different sampling times (i min).

The first-order rate equation of Lagergren, which is one of the most widely used equations for the sorption of solute from a liquid solution, was employed to describe the pigment sorption kinetics.

Rearrangement of the Lagergren model [28] was used for the variation of the adsorbed pigment as a function of time:

$$\frac{q_t}{q_r} = 1 - \frac{exp[(-kt_r)(t/t_r)]}{1 - exp(-kt_r)} \tag{2}$$

where q_t and q_r are the amount of adsorbed pigment at time t (min) and at equilibrium (%), respectively, k is the first-order rate constant (min^{-1}), and t_r is the longest time of the sorption process (min).

2.7. Data Analysis

The model parameters were estimated by non-linear regression analysis [29]. Results were analysed statistically by factorial ANOVA to test statistical differences ($p < 0.05$), followed by Tukey's test at 5% probability for comparisons. Statistical and regression analyses were made with STATISTICA 7.1. (StatSoft, Inc., Tulsa, OK, USA, 2005,).

3. Results and Discussion

Dying Properties of Pigments

The effect of the mordant used on L^*, a^*, and b^* parameters are given in Table 2; L^* represents a lightness value (a higher lightness value represents a lower colour yield). a^* and b^* represent the tone of the colour; positive values of a^* and b^* represent redder and yellowish tones, respectively.

Total colour difference ΔE^* between the dyed fabrics and the undyed wool was calculated as:

$$\Delta E^* = \sqrt{\left(L^*_U - L^*_D\right)^2 + \left(a^*_U - a^*_D\right)^2 + \left(b^*_U - b^*_D\right)^2} \tag{3}$$

where sub-index U and D represent undyed and dyed wool, respectively, for L^*, a^*, and b^* values.

Fabrics were qualitatively perceived as different; the dyeing process using mordant F presented a red colour while the dyeing process using mordant A showed a red tending to brown colour. Colorimetric studies indicated that the tested mordants significantly affected the colour exhibited by the fabric. The dyed wool using mordant F presented a higher total colour difference from the undyed wool than the dyed wool using mordant A (Table 2). Higher values of a^* were obtained with mordant F in comparison with mordant A, indicating that more reddish tones can be obtained using mordant F. Meanwhile, b^* values indicated that more yellowish tones can be obtained with mordant A. These results were more evident with the hue values obtained; hue values near to 0 indicated the degree of redness while values near to 90 represented the level of yellowness. Moreover, the wool dyed using mordant F presented a higher saturation of colour (Chroma) and a higher yield (L^*) than that dyed with mordant A.

Dyed fabrics obtained here presented a stronger shade (red colour) than the optimum conditions reported for the dyeing of cotton using pigments produced by different fungal strains (*Monascus purpureus*, *Isaria farinose*, *Emericella nidulans*, *Fusarium verticilliodes* and *Penicillium purpurogenum*) [12].

Similarly, a^* and b^* values were higher than those reported for the dyeing of leather and silk with *Talaromyces* pigments [11,30]; however, those results were attained without the addition of mordants.

Mordant concentration also affected the colour exhibited by the dyed fabric. When mordant F was used, the colour yield obtained increased with concentration. However, there was not a statistical difference between F2 and F3. It is noted that a more reddish colour (a^*) was obtained by increasing the mordant concentration while there was not a statistical difference achieving a yellow tone (b^*) between F2 and F3 or between F1 and F2. In terms of colour saturation, hue, and total colour difference, there was not statistical difference between F2 and F3 indicating that if an F concentration between 20 and 30% *w/w* is used during the mordanting process, the same colour can be obtained.

Different results were obtained using mordant A; colour yield increased by increasing mordant concentration. A more reddish colour (a^*) was obtained using the highest concentration while the yellowish tone (b^*) was not affected by concentration. Furthermore, the concentration did not affect colour saturation (Chroma). A higher degree of redness (Hue) and higher total colour difference (ΔE^*) was attained using the highest concentration of mordant A.

These results are consistent with those reported by Arroyo-Figueroa et al. (2011) [26], who stated that a colour variation in terms of CIELAB scale is obtained as function of mordant concentration during the dyeing process using natural colourants.

In technical dyeing using synthetic dyes, a total colour difference of 1 is accepted as a tolerable colour difference between dyeings. However, with the introduction of natural dyes into the textile dyeing process, a wider total colour difference can be accepted ($\Delta E^* = 2$) [31]. Results demonstrated that, independent of the mordant and concentration tested, an acceptable colour difference between dyeings was obtained ($\Delta E^* = 0.71$–1.70).

The effect of the dyeing process time on pigment uptake using different mordants at different concentrations is shown in Figure 1. It can be seen that dyeings with mordant F presented higher values of pigment uptake (42.54–82.44%) than mordant A (33.98–41.98%). In Table 3 are listed the kinetic parameters obtained with the regression analysis of Equation (2). When mordant F was used, pigment uptake at equilibrium (q_r) increased as mordant concentration increased. Mordant A did not show a significant difference between concentrations A1 and A2, reaching the maximum pigment uptake with concentration A3. The higher pigment uptake here obtained ($81.33 \pm 0.43\%$) is similar to the maximum (80%) reported by Velmurugan et al. (2010) [12], which was achieved dyeing cotton with red pigments produced by *Monascus purpureus*.

Table 2. L^*, a^*, b^* values of dyed wool.

Experiment	Colour Coordinates					
	L^*	a^*	b^*	Chroma	Hue	ΔE^*
Wool	93.89 ± 1.16	0.44 ± 0.22	5.15 ± 1.47	5.17 ± 1.47	84.94 ± 2.63	0.00
F1	31.39 ± 1.11 [b]	24.35 ± 0.44 [c]	21.01 ± 0.20 [c]	32.16 ± 0.23 [b]	40.79 ± 0.76 [b]	68.78 ± 0.96 [b]
F2	28.20 ± 0.64 [a]	27.62 ± 0.68 [b]	20.61 ± 0.41 [bc]	34.47 ± 0.64 [a]	36.73 ± 0.78 [a]	72.76 ± 1.70 [a]
F3	27.53 ± 0.65 [a]	28.91 ± 0.35 [a]	21.87 ± 0.81 [b]	36.25 ± 0.25 [a]	37.10 ± 1.34 [a]	74.14 ± 0.90 [a]
A1	44.47 ± 0.68 [e]	14.84 ± 0.15 [e]	27.34 ± 0.32 [a]	31.11 ± 0.34 [d]	61.50 ± 0.21 [d]	56.07 ± 0.71 [d]
A2	42.48 ± 0.43 [d]	15.07 ± 0.24 [e]	27.83 ± 0.25 [a]	31.65 ± 0.13 [d]	61.56 ± 0.57 [d]	58.09 ± 1.65 [d]
A3	37.08 ± 0.56 [c]	17.14 ± 0.35 [d]	27.76 ± 0.14 [a]	32.63 ± 0.18 [d]	58.32 ± 0.57 [c]	63.40 ± 1.52 [c]

Different letters in each column indicate significant differences (Tukey's post-hoc comparison, $p < 0.05$).

Results showed that there was no statistical difference in terms of exponential rate (k) using mordant F at all concentrations and mordant A at concentrations A2 and A3. Only mordant A showed the lowest exponential rate at the lowest concentration studied.

Table 3. Kinetic parameters and goodness of fit of the data.

Experiment	q_r, %	k, min^{-1}	R^2
F1	50.56 ± 0.57 [c]	0.0504 ± 0.0028 [a]	0.98
F2	76.99 ± 0.55 [b]	0.0542 ± 0.0021 [a]	0.99
F3	81.33 ± 0.43 [a]	0.0537 ± 0.0015 [a]	0.99
A1	36.91 ± 0.82 [e]	0.0218 ± 0.0011 [b]	0.98
A2	35.40 ± 0.33 [e]	0.0497 ± 0.0023 [a]	0.99
A3	41.02 ± 0.39 [d]	0.0569 ± 0.0030 [a]	0.99

Different letters in each column indicate significant differences (Tukey's post-hoc comparison, $p < 0.05$).

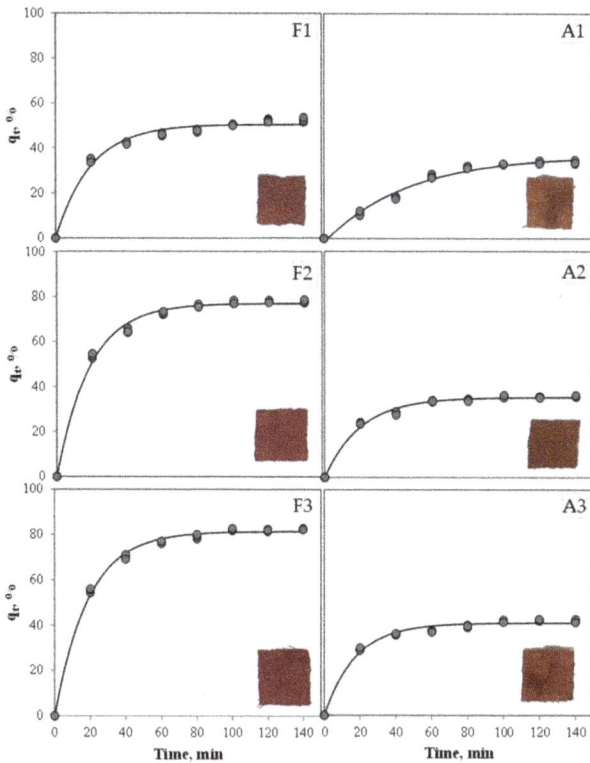

Figure 1. Effect of mordant type and concentration on pigment sorption on wool. Circles represent experimental data points and continuous lines represent the Lagergren model.

In literature, sorption curves are classified into four zones (0 to 4) according to the kt_r value; the kinetic behavior goes from no adsorption to drastic (Please see sorption kinetic behavior reported by Tseng et al. [28].

The time of half dyeing $t_{1/2}$ is also used to express the time required for a fabric to adsorb half of the amount of the dye adsorbed at equilibrium ($q_t/q_r = 0.5$), calculated as:

$$t_{1/2} = \frac{ln(2)}{k} \tag{4}$$

Table 4 lists the dimensionless rate constants (kt_r), the half dyeing time values ($t_{1/2}$), and the zone of kinetic sorption behavior presented for the mordants studied at three concentration levels.

Table 4. Sorption kinetic parameter (kt_r), $t_{1/2}$ and zone of the sorption behavior.

Experiment	kt_r	$t_{1/2}$, min	Zone
F1	7.06	13.74	III
F2	7.58	12.79	III
F3	7.51	12.92	III
A1	3.04	31.84	II
A2	6.95	13.95	III
A3	7.97	12.17	III

Results showed that most of sorption kinetic curves fell into zone III (Kinetic curve type, rapidly rise; kinetic behavior, quick), while only the dyeing process using A1 fell into zone II (Kinetic curve type, continuously rising; kinetic behavior, good).

The $t_{1/2}$ values obtained were in range with those reported for the optimised conditions for dying wool ($t_{1/2}$ = 13.1 min) with natural anthraquinones dyes produced by the fungus *Fusarium oxysporum* [32]. The $t_{1/2}$ values are relatively short in comparison with the longest time of the sorption process (t_r), which could result in undesired colour yields. Thus, the above kinetics are useful to establish a target value (q_t/q_r > 0.5) to assist the engineering design and optimisation of the dyeing process of wool by *Talaromyces* pigments.

4. Conclusions

The above results showed the feasible potential applications of the pigments produced by *Talaromyces* spp.

These pigments could serve as a useful alternative source for the natural dyeing of wool textiles. High values of pigment uptake on the fabric were obtained. The values of dyeing rate constant (k), half-time of dyeing ($t_{1/2}$) and sorption kinetics behaviour compare well with other natural dyes used for dyeing textile. However, the process still needs to be optimised in terms of dying process conditions (pigment concentration, pH, temperature).

It is concluded that the mordant used plays an important role in the dying process. Colorimetric studies showed that it is possible to attain a strong red shade; therefore, these pigments can compete with other synthetic or natural red dyes.

From this point onward, full characterisation of the molecules, toxicity studies, and more in-depth analyses such as molecular interactions between the dye and the fabric are worth tfurther investigation.

Acknowledgments: Author Lourdes Morales-Oyervides acknowledges CONACyT-México for the financial support provided for conducting her PhD studies (215490-2011).

Author Contributions: Julio Cesar Montañez and Jorge Oliveira conceived and designed the experiments; Lourdes Morales-Oyervides performed the experiments; Jorge Oliveira and Maria Sousa-Gallagher analysed the data; Alejandro Méndez-Zavala and Julio Cesar Montañez contributed reagents/materials/analysis tools; Lourdes Morales-Oyervides wrote the paper.

Conflicts of Interest: The authors declare no conflict of interest.

References

1. Carocho, M.; Morales, P.; Ferreira, I.C.F.R. Natural food additives: Quo vadis? *Trends Food Sci. Technol.* **2015**, *45*, 284–295. [CrossRef]
2. Shahid, M.; Mohammad, F. Recent advancements in natural dye applications: A review. *J. Clean. Prod.* **2013**, *53*, 310–331. [CrossRef]
3. Fernández-García, E.; Carvajal-Lérida, I.; Jarén-Galán, M.; Garrido-Fernández, J.; Pérez-Gálvez, A.; Hornero-Méndez, D. Carotenoids bioavailability from foods: From plant pigments to efficient biological activities. *Food Res. Int.* **2012**, *46*, 438–450. [CrossRef]
4. Kostick, R.; Wang, S.; Wang, J. Color Cosmetics—Coloring/Staining the Skin from Fruit and Plant Color Pigments. Patent US20060280762, 2006.
5. Venil, C.K.; Zakaria, Z.A.; Ahmad, W.A. Bacterial pigments and their applications. *Process Biochem.* **2013**, *48*, 1065–1079. [CrossRef]
6. Arai, T.; Kojima, R.; Motegi, Y.; Kato, J.; Kasumi, T.; Ogihara, J. PP-O and PP-V, *Monascus* pigment homologs, production and phylogenetic analysis in *Penicillium purpurogenum*. *Fungal Biol.* **2015**, *119*, 1226–1236. [CrossRef] [PubMed]
7. Mapari, S.A.S.; Hansen, M.E.; Meyer, A.S.; Thrane, U. Computerized screening for novel producers of *Monascus*-like food pigments in *Penicillium* species. *J. Agric. Food Chem.* **2008**, *56*, 9981–9989. [CrossRef] [PubMed]

8. Santos-Ebinuma, V.C.; Roberto, I.C.; Teixeira, M.F.S.; Pessoa, A., Jr. Improvement of submerged culture conditions to produce colorants by *Penicillium purpurogenum*. *Braz. J. Microbiol.* **2014**, *45*, 731–742. [CrossRef] [PubMed]

9. Morales-Oyervides, L.; Oliveira, J.C.; Sousa-Gallagher, M.J.; Méndez-Zavala, A.; Montañez, J.C. Selection of best conditions of inoculum preparation for optimum performance of the pigment production process by *Talaromyces* spp. using the Taguchi method. *Biotechnol. Prog.* **2017**. [CrossRef] [PubMed]

10. Padmapriya, C.; Murugesan, R. Optimization of physical parameters on polyketide red pigment production from *Penicillium purpurogenum*. *Trends Biosci.* **2014**, *7*, 1664–1667.

11. Sudha; Gupta, C.; Aggarwal, S. Optimization and extraction of extra and intracellular color from *Penicillium minioluteum* for application on protein fibers. *Fibers Polym.* **2017**, *18*, 741–748. [CrossRef]

12. Velmurugan, P.; Kim, M.J.; Park, J.S.; Karthikeyan, K.; Lakshmanaperumalsamy, P.; Lee, K.J.; Park, Y.J.; Oh, B.T. Dyeing of cotton yarn with five water soluble fungal pigments obtained from five fungi. *Fibers Polym.* **2010**, *11*, 598–605. [CrossRef]

13. Dhale, M.A.; Vijay-Raj, A.S. Pigment and amylase production in *Penicillium* sp. NIOM-02 and its radical scavenging activity. *Int. J. Food Sci. Technol.* **2009**, *44*, 2424–2430. [CrossRef]

14. Morales-Oyervides, L.; Oliveira, J.C.; Sousa-Gallagher, M.J.; Méndez-Zavala, A.; Montañez, J.C. Effect of heat exposure on the colour intensity of red pigments produced by *Penicillium purpurogenum* GH2. *J. Food Eng.* **2015**, *164*, 21–29. [CrossRef]

15. Arora, S. Textile Dyes: It's impact on environment and its treatment. *J. Bioremediation Biodegrad.* **2014**, *5*, 1. [CrossRef]

16. Borges, M.E.; Tejera, R.L.; Díaz, L.; Esparza, P.; Ibáñez, E. Natural dyes extraction from cochineal (*Dactylopius coccus*). New extraction methods. *Food Chem.* **2012**, *132*, 1855–1860. [CrossRef]

17. Patil, S.; Sivanandhan, G.; Thakare, D. Effect of physical and chemical parameters on the production of Red exopigment from *Penicillium purpurogenum* isolated from spoilt onion and study of its antimicrobial activity. *Int. J. Curr. Microbiol. Appl. Sci.* **2015**, *4*, 599–609.

18. Chadni, Z.; Rahaman, M.H.; Jerin, I.; Hoque, K.M.; Reza, M.A. Extraction and optimisation of red pigment production as secondary metabolites from *Talaromyces verruculosus* and its potential use in textile industries. *Mycology* **2017**, *8*, 48–57. [CrossRef]

19. Sopandi, T.; Wardah, W. Sub-Acute toxicity of pigment derived from *Penicillium resticulosum* in mice. *Microbiol. Indones.* **2012**, *6*, 35–41. [CrossRef]

20. Cruz-Hernández, M.; Rodráguez, R. Aislamiento y caracterización morfológica de cepas microbianas degradadotas de taninos. In Proceedings of the XXII Annual Meeting of the Mexican Academy of Chemical Engineering (AMIDIQ), Mazatlan, Mexico, 9 September 2001.

21. Espinoza-Hernández, T. Caracterización Morfológica, Fisiológica y Molecular de tres Cepas Fúngicas Productoras de Pigmentos. Bachelor's Thesis, Universidad Autónoma de Coahuila, Saltillo, México, 2004.

22. Frisvad, J.C.; Yilmaz, N.; Thrane, U.; Rasmussen, K.B.; Houbraken, J.; Samson, R.A. Talaromyces atroroseus, a new species efficiently producing industrially relevant red pigments. *PLoS ONE* **2013**, *8*, e84102. [CrossRef] [PubMed]

23. Morales-Oyervides, L.; Oliveira, J.C.; Sousa-Gallagher, M.J.; Méndez-Zavala, A.; Montañez, J.C. Quantitative assessment of the impact of the type of inoculum on the kinetics of cell growth, substrate consumption and pigment productivity by *Penicillium purpurogenum* GH2 in liquid culture with an integrated stochastic approach. *Food Bioprod. Process.* **2015**, *96*, 221–231. [CrossRef]

24. Méndez-Zavala, A. Identificación de Factores que Afectan la Producción de Pigmentos por Penicillium purpurogenum GH2 y Obtención de Perfiles Cromatográficos. Master's Thesis, Universidad Autónoma de Coahuila, Saltillo, México, 2011.

25. Méndez-Zavala, A.; Pérez, C.; Montañez, J.C.; Martínez, G.; Aguilar, C.N. Red pigment production by *Penicillium purpurogenum* GH2 is influenced by pH and temperature. *J. Zhejiang Univ. Sci. B* **2011**, *12*, 961–968. [CrossRef] [PubMed]

26. Arroyo-Figueroa, G.; Ruiz-Aguilar, G.M.L.; Cuevas-Rodriguez, G.; Sanchez, G.G. Cotton fabric dyeing with cochineal extract: Influence of mordant concentration. *Color. Technol.* **2011**, *127*, 39–46. [CrossRef]

27. Velmurugan, P. Studies on the Production and Dyeing Properties of Water Soluble Pigments from Filamentous Fungi. Ph.D. Thesis, Bharathiar University, Coimbatore, India, 2008.

28. Tseng, R.-L.; Wu, F.-C.; Juang, R.-S. Characteristics and applications of the Lagergren's first-order equation for adsorption kinetics. *J. Taiwan Inst. Chem. Eng.* **2010**, *41*, 661–669. [CrossRef]

29. Hill, C.G.J.; Grieger-Block, R.A. Kinetic data: Generation, interpretation, and use. *Food Technol.* **1980**, *34*, 55–66.

30. Gupta, C.S.; Aggarwal, S. Dyeing wet blue goat nappa skin with a microbial colorant obtained from *Penicillium minioluteum*. *J. Clean. Prod.* **2016**, *127*, 585–590.

31. Bechtold, T.; Mahmudali, A.; Mussak, R. Natural dyes for textile dyeing: A comparison of methods to assess the quality of Canadian golden rod plant material. *Dyes Pigments* **2007**, *75*, 287–293. [CrossRef]

32. Nagia, F.A.; El-Mohamedy, R.S.R. Dyeing of wool with natural anthraquinone dyes from *Fusarium oxysporum*. *Dyes Pigment.* **2007**, *75*, 550–555. [CrossRef]

MDPI AG

St. Alban-Anlage 66

4052 Basel, Switzerland

Tel. +41 61 683 77 34

Fax +41 61 302 89 18

http://www.mdpi.com

Journal of Fungi Editorial Office

E-mail: jof@mdpi.com

http://www.mdpi.com/journal/jof